I0817605

Big Trust

BIG TRUST

Rewire Self-Doubt, Find Your Confidence, and Fuel Success

DR. SHADÉ ZAHRAI

with Fayçal Sekkouah

HARPERONE

An Imprint of HarperCollins*Publishers*

Without limiting the exclusive rights of any author, contributor or the publisher of this publication, any unauthorized use of this publication to train generative artificial intelligence (AI) technologies is expressly prohibited. HarperCollins also exercise their rights under Article 4(3) of the Digital Single Market Directive 2019/790 and expressly reserve this publication from the text and data mining exception.

BIG TRUST. Copyright © 2026 by Shadé Zahrai and Philip Faysal Sekkouah. All rights reserved. No part of this book may be used or reproduced in any manner whatsoever without written permission except in the case of brief quotations embodied in critical articles and reviews. For information, address HarperCollins Publishers, 195 Broadway, New York, NY 10007. In Europe, HarperCollins Publishers, Macken House, 39/40 Mayor Street Upper, Dublin 1, D01 C9W8, Ireland.

HarperCollins books may be purchased for educational, business, or sales promotional use. For information, please email the Special Markets Department at SPsales@harpercollins.com.

harpercollins.com

FIRST EDITION

Designed by Jason Kayser
Illustrations by Maya P. Lim

Library of Congress Cataloging-in-Publication Data has been applied for.

ISBN 978-0-06-337815-5
ISBN 978-0-06-349010-9 (international edition)

PRINTED IN THE UNITED STATES OF AMERICA

25 26 27 28 29 LBC 6 5 4 3 2

For our parents . . .
Thank you for your endless love, belief, prayers, and sacrifice.
Everything we are began with you.

And to you, dear followers . . .
Without you, this book would never have come into being.
May it serve you well.

"Be calm,
be strong,
be grateful,
and become a lamp full of light. . . ."

— 'Abdu'l-Bahá[1]

CONTENTS

The Third Attribute: Autonomy

The Fourth Attribute: Adaptability

The Path to Big Trust

INTRODUCTION

We Are All More Than Our Self-Doubts

It's 10 p.m. I'm lying in bed, staring at the ceiling, my thoughts racing like they always do at the exact moment I'm supposed to be winding down for sleep. I know better than to grab my phone, but I do it anyway, fumbling in the dark. Airplane mode off. Brightness down.

Earlier that day, I'd been working on a program about accelerating career growth. The topics sounded great on paper, but something about it felt . . . off.

For years, I'd shared frameworks, strategies, and motivational fuel that lit people up and helped them take action. Yet I was midway through my work with this group of mid-career professionals, and I had a gnawing sense that my approach was missing the mark, that it would ignore their deeper frustrations and worries.

Across the board, I knew they weren't just stuck in their careers; they were stuck in *themselves*. And honestly, I'd been there too. The grind without the growth. The effort without the results. That maddening sense of pouring yourself into something and still feeling like it's not enough.

This sinking feeling was reminding me of one of my earliest encounters with self-doubt. I was nineteen, sitting in a Legal Jurisprudence class at Macquarie University in Sydney, Australia, and everything about it felt *wrong*. I had chosen law because I had the grades and because everyone said, "Don't waste them." My parents, my teachers, my immigrant grandparents who had fled Iran to build a life in the West[1]—they all had visions of "lawyer" or "doctor" for my future. To balance things out, I paired law with a psychology degree, something that actually made sense to me. Human behavior had always fascinated me.

But on that first day of university, as I sat in class, law felt foreign—like a mistake. No, *I* felt like the mistake. My classmates tossed out answers while I scrambled to keep up, barely grasping the jargon, let alone the concepts. Heat rushed to my face. The insecurities I hadn't realized were lurking beneath the surface roared to life:

I don't belong here.
I'm not smart enough.
I'll never be able to keep up.

The moment class ended, I practically ran to the department advisor's office. The words spilled out before I could stop them: "I don't understand anything—I need to drop law."

I expected encouragement. Reassurance. Someone to remind me it was just the first day, that I'd find my footing soon enough. Maybe even that I was smarter than I thought. I didn't admit it to myself, but I had wanted the advisor to take my self-doubts away, to "fix" things for me. What I got instead was silence. Then she handed me a *course withdrawal* form.

It felt like a punch to the gut. Like she saw someone who was not worth believing in.

I didn't drop law, though. I walked out of her office feeling humiliated but also strangely defiant. I didn't want her (or anyone) to see me as someone who couldn't handle it. So I pushed myself. I worked harder. Stayed up later. Powered through moments of doubt, convinced that achievement would eventually silence the insecurities.

Of course, it didn't. (As I've since learned, so many others try, and fail, to do the same.)

Self-doubt followed me from law school into corporate law, where everything felt *so* hard. It followed me into my banking career, where I thought a career change might leave the doubt behind.

It didn't.

It shadowed me in meetings, silencing ideas I wanted to share. It was the anxious churn in my stomach, the overthinking that ruined my sleep, the constant second-guessing that never let up.

Even when opportunities came my way, I rationalized my way out of them. *I'm not ready. The timing's not right.* I told myself I was being cautious and wise. But really, it was fear. Fear wrapped in overthinking. Fear moonlighting as prudence. And still, I pushed myself, trying to drown out doubts that refused to let go.

That night, as I laid in bed all those years later, eyes locked on the faint glow of my phone screen, those memories resurfaced.

Then, it hit me. What if I've been asking the wrong question all along?

Maybe it's not about how to go faster or push harder. Maybe the real question is: *Why am I stuck in the first place?*

If I could figure out what was holding me back—what was *really* in the way—I knew I could help others move forward with their own struggles in a way that felt lighter. Less forced. Instead of feeling panic or dread that there was something missing in my work, I felt something else. Relief. Clarity. Possibility. Maybe there *was* another way, and I was ready to find it.

The Stickiness of Self-Doubt

Over the years, I learned, on a personal level, how to push past my own self-doubt. One step at a time, I rebuilt my career, strengthened my confidence, and achieved milestones I once thought were out of reach. I went back to university, earned credentials, and became a peak performance specialist, behavioral strategist, executive coach, and researcher, building a career in helping others unlock their potential.

But it occurred to me that maybe my perspective was too narrow.

For years, my work (and my own path) had focused heavily on the *visible* markers of success: landing the promotion, hitting the next career milestone, achieving a major life goal. These were the outcomes people sought, the things that felt like the solution to their self-doubt. And I was helping them get there. And yet, even when clients reached their goals, something was missing. Their self-doubt didn't lift (and neither did mine). If anything, it evolved, morphing into new fears, new insecurities, new thresholds of "not enough."

So, in the quiet of that night, I tapped out a question on my phone. A simple, provocative thought that I sent to a list of a few hundred psychologists, performance coaches, and professors:

"If you wanted to sabotage someone's success and happiness, what's the most effective way to do it?"

By morning, forty-eight responses had poured in. They were blunt, unfiltered, and eerily similar, saying things like:

"Make them overthink until they're paralyzed by indecision."

"Plant seeds of self-doubt so they question every move they make."

"Get them to compare themselves with others and feel like they're never enough."

"Constantly remind them of their weaknesses so they lose faith in themselves."

"Convince them they're insignificant, so they shrink away from their potential."

Different words, same conclusion: To sabotage anyone's potential, cultivate self-doubt. It doesn't take grand, elaborate schemes. Just plant a tiny seed of self-doubt, and it will take root—it will stick. A person's mind will handle the rest.

Let me tell you a story that demonstrates how self-doubt sticks around. At the end of 2021, my husband, Fayçal, and I rescued a stray dog. We named her Bonbon—French for "candy," because she's as sweet as they come. Plus, as French is Fayçal's first language, the name just felt right. Bonbon quickly became part of our world. And so did an unexpected daily ritual.

Every walk through the fields near our home left us covered in burrs. These tiny, seeded pods were absolute nuisances. They'd become entangled in Bonbon's fur, my socks, even my hair. Every day, without fail, we'd sit and pick them out, one by one.

Turns out, this wasn't just our problem.

Back in 1941, a Swiss engineer named George de Mestral went hiking with his dog and came home covered in burrs too. But instead of just being annoyed, he got curious. Under a microscope, he discovered that these annoying little things had tiny, barbed hooks that latched onto anything they touched. His eureka moment led to what we now know as Velcro.

Here's the thing about burrs: They don't just stick. They *cling*. Relentlessly. Stubbornly. Almost impossibly. And so does self-doubt.

It starts small. A passing thought, something you could easily brush off. Then it hooks in. It tangles its way into your decisions, your confidence, even the way you see yourself. Left unchecked, self-doubt becomes part of your mental operating system. It runs quietly in the background, shaping your choices without you even realizing it. This is where self-doubt turns toxic. Instead of keeping you vigilant, it makes you second-guess *everything*.

Like my clients did.

Like I did.

Like I'm guessing you do.

The mistake isn't having self-doubt. The mistake is thinking you have to defeat it.

You don't overcome self-doubt by silencing it—though that's what too many of us believe (and too many "experts" teach). You take away its power by noticing it, understanding it, and changing your relationship to it. You unhook from self-doubt when you see it for what it is: a misguided attempt to protect you. That's the irony. Self-doubt isn't trying to *sabotage* you; it's trying to *shield* you.

When channeled the right way, self-doubt can be useful. It's like a friend in the passenger seat. The "good" kind leans over and says, "Hey, your fuel light just came on—better find a gas station soon." The "bad" kind is the obnoxious backseat driver who won't shut up, insisting you take the next exit, even though your GPS is clearly right. The longer you listen to that voice in the backseat, the more convincing it becomes. *Maybe I really am lost. Maybe I shouldn't be driving at all.*

The truth is that the world we live in today is perfectly engineered to amplify self-doubt. We're more exposed than ever—scrolling through curated lives, juggling blurred lines between public and private selves, bombarded with messages about "making it" and endlessly upgrading ourselves. It's no wonder self-acceptance feels almost . . . impossible.

More success doesn't erase the doubt. More knowledge doesn't silence it. And despite all our modern tools, psychological insights, and endless "life hacks," self-doubt isn't getting easier to manage; it's getting *louder*. And there's a reason why. Self-doubt isn't something you can block out or ignore. It's something you have to acknowledge. What I've learned—from my own life, from the research, and from the thousands of people I've worked with—is that, understood correctly, listened to well, and transformed wisely, self-doubt can be one of your greatest tools for growth.

From Self-Doubt to Big Trust

True success and peak performance—what I call operating from your "peak state"—starts with self-trust. It's the ability to show up fully, stay grounded under pressure, and back yourself when it counts.

Self-doubt, though, is the ultimate kryptonite of that trust. It annihilates your belief in yourself faster than anything else.

For some people, like Sadia, it keeps them from even trying. When I met her, she'd been doing the work of a leader for years—running projects, making decisions, and mentoring her peers—but without the title. When her mentor encouraged her to apply for a promotion, she should have felt ready. Instead, she hesitated. *What if I apply and don't get it? Or worse, get it and fail?*

I asked her, "If someone else with your exact experience were in your position, what would you tell them?"

Without thinking, she replied, "That they should absolutely go for it."

That's one way self-doubt works. It distorts reality. It feels true, but truth and perception aren't always the same thing.

Then there's the version of self-doubt that doesn't stop you but keeps you running. That was Ray. On paper, he was the definition of success: top of his class, a series of fast-tracked promotions. Yet he was miserable. Everything felt harder than it needed to be, and nothing ever felt truly satisfying. When I asked him what he thought the reason might be, he then admitted, "I keep thinking the *next* win will be the one that finally makes me feel like I've made it." For years, he believed that once he hit a certain salary, got a certain title, or proved himself to the right people, he'd stop second-guessing himself and finally feel at peace. But the feeling never arrived. It just attached itself to the *next* milestone.

"I don't get it," he admitted. "I've done *everything* I was supposed to. Why do I feel this way?"

For both Sadia and Ray, they had built their careers on different responses to self-doubt—Sadia by holding back, Ray by pushing harder—but underneath, the same truth remained: Neither fully trusted themselves.

I see this all the time with executive coaching clients. They're smart and highly capable. But the mental strain of their self-doubt bleeds into *everything*. They toss and turn at night, wake up exhausted, and start the next day on caffeine and fumes. And over time, the very thing they doubted—their value, their ability to contribute—becomes a self-fulfilling prophecy. Not because they aren't capable or lack talent, but because self-doubt drained them of the confidence and energy to perform at their best.

But when you navigate self-doubt differently, prepare for it, and even use it as a tool for self-reflection and growth, things change. You stop hesitating. You worry less. You manage your anxiety. You sleep better. You wake up clearer, calmer, more focused. And suddenly, you're healthier, happier, and way more productive.

Your self-trust grows and serves as the ultimate cheat code. And over time, that trust solidifies into something even deeper: Big Trust. That is, the ability to move through life freely, without being hijacked by doubt. You back yourself when it counts.

Once Sadia understood that her fear wasn't about incompetence but driven by an inability to trust her own skills, she started making small shifts. She leaned into the strengths she *did* trust, like her ability to mentor others, her problem-solving skills, and her deep knowledge of the work she had been doing for years. She applied for and got the promotion. Months later, she told me, "I still don't know everything. But I'm realizing no one does."

Ray, on the other hand, *knew* he was competent, but he was coming to realize that his sense of self-worth hinged entirely on his achievements. If he wasn't ranking near the top of the sales leaderboard, exceeding his quarterly targets, or receiving positive feedback from the leadership team, he felt worthless. For him, building self-trust meant separating his identity from his accomplishments. Some days, that meant pausing to acknowledge his progress before moving the goalposts. Other days, it was simply reminding himself: *Success isn't just numbers. It's impact. A well-timed insight, helping a colleague, earning trust.* Ray started tracking *those* wins, not just the bottom line.

My Obsession with Self-Doubt: Developing a New Framework

For as long as I can remember, I've been the one people come to with their doubts, fears, and insecurities. It started early. In middle school, I somehow became the unofficial keeper of "secrets." By high school, I was the designated advice-giver (though I'm not quite sure why, given I had just as little life experience as they did). As the only daughter of Persian immigrant parents, I was the one my mom would confide in before my parents separated when I was fifteen.

By university, and later at work, I found myself in the familiar role of listener and helper—classmates, colleagues, and people I had barely met sought me out for advice. It was like I had "Here to Help" tattooed on my forehead. And honestly? I loved it.

Years later, I turned that role into a career. I got an MBA, trained as an executive coach, and pursued a PhD that spans personality and organizational behavior (where much of this book's model was born). My obsession is in uncovering what drives people—and what stops them in their tracks.

In 2017, Fayçal and I launched a global leadership development company focused on helping people break through their internal roadblocks. That mission turned into an ongoing experiment—analyzing the ways self-doubt limits human potential and, more importantly, figuring out how to break the cycle.

Through facilitating high-stakes corporate conversations, coaching teams, and training thousands of leaders from Fortune 500 giants like Microsoft, J.P. Morgan, Deloitte, and LVMH, I've spent years turning science, psychology, and real-world data into actionable tools that work. We help clients unlock greater creativity, enthusiasm, and peak performance.

Of course, I haven't done this alone. Fayçal brings his own hard-earned wisdom. With his razor-sharp ability to cut to the heart of any problem and a knack for infusing optimism into every situation, he's been the perfect complement to my deep-dive obsession with human behavior. A serial entrepreneur and investor, he's launched world-first consumer electronics, mentored at the Branson Centre of Entrepreneurship, and consulted at the G20. He knows what it means to take risks, face fear head-on, and make things happen.

Together, we started our business at a small scale. And then . . . the world changed.

When COVID-19 hit, suddenly the entire world plunged into a collective crisis of self-doubt. Careers, relationships, and entire futures suddenly teetered. People everywhere questioned their decisions, second-guessed themselves, and felt stuck in ways they never had before. We pivoted quickly, moving to virtual training and sharing educational videos on social media. Remarkably, they went viral. Thousands of people reached out from around the globe. Their messages all contained the same thread running through them: People were feeling swallowed whole by self-doubt that was being amplified by the chaos around them. They didn't know how to stop it.

That's when I knew it was time to dissect self-doubt differently—to consider it not just as a frustrating mental loop, but as a pattern we could actively dismantle or transform. The more I studied self-doubt—and, more importantly, began to practice new techniques on my own fears and uncertainties—the more I saw just how powerful this work is.

This project led directly to my PhD research, where I discovered fascinating truths about our relationship with self-doubt. Through analyzing the data and cross-referencing it with years of client notes, I found that those who struggle most with self-doubt tend to be hyper-fixated on it. They *become* the self-doubt. They don't just question their abilities or worth; they doubt *who* they are at their core. It shapes their very identity and can feel like a full-on crisis of self. The harder people try to "fix" themselves, the more they amplify all of their insecurities.

Through my research I also discovered that self-doubt isn't just one giant, messy "blob" of worry, fear, hesitation, and stress. It's the combination of four distinct, trainable Attributes that shape your self-trust—and your identity. Together, they form your individual Doubt Profile.

But here's what's most exciting: These Attributes aren't fixed traits. Each one is a pattern of thought and behavior, repeated over time, often without you even realizing it. In that way, they reflect your habits. And like any habit, they can be changed. If you strengthen the *right* habits aligned with each of the Four Attributes, you don't just reduce self-doubt. You fundamentally change *who you are.* That's how Big Trust is built—not all at once, but one small habit at a time.

Here's how the Four Attributes of your Doubt Profile break down:

1. **Acceptance** — Do you believe you're enough as you are?
2. **Agency** — Do you trust your skills and abilities?
3. **Autonomy** — Do you feel that you can shape your path?
4. **Adaptability** — Can you stay emotionally grounded when doubt arises?

Think of yourself as George de Mestral, curiously examining those burrs and their tiny, hooked barbs under a microscope. By breaking self-doubt into its "hooks"—the four core Attributes of Acceptance, Agency, Autonomy, and Adaptability—you can pinpoint exactly where it's latching on and you can begin to unhook yourself from its grip, one pattern and one habit at a time. That's how you build a stronger sense of self and unlock more confidence, courage, and energy for the things that truly matter.

How to Make This Book Stick

This book is a guide for real life, grounded in science, and designed to get results. It's not just a *thinking* book; it's a *doing* book. Here's what you'll find inside:

- **Insights, inspiration, and motivation to help you interrupt, challenge, and reimagine your beliefs.** Your beliefs about yourself are shaped by repeated thought patterns. But those thoughts are not facts—and they can change. As you shift your thoughts and behaviors, you begin forming new habits—ones that reinforce trust in yourself rather than doubt. Through client stories (with names and identifying details changed for privacy), real-world scenarios, insights from well-known figures, and stories from my own life, you'll see exactly how doubt operates—and how those who've overcome it have done so. Once you've seen all that, you can do it too.
- **Practical, specific guidance on working with the Attributes of Doubt—quieting the voices that hold you back while strengthening the qualities that naturally counteract them.**

With Acceptance, you'll anchor into your worth. With Agency, you'll rebuild belief in your abilities and what you're capable of. With Autonomy, you'll feel more empowered and in control of your choices. And with Adaptability, you'll handle challenges and emotions without falling apart. When practiced consistently, they unlock more confidence, courage, and energy for the moments that *actually* matter.

- **The self-trust tools.** You'll acquire a toolkit to help you build self-trust, one small step at a time:

    The Practices—These are a mix of reflections to take stock of where you are, simple techniques to challenge self-doubt in the moment, and exercises to help you unhook from doubt at a deeper level. Each practice helps reinforce ways of thinking and habits linked to the Attribute you're developing. An entire Attribute of emotional Adaptability will help you understand how emotions amplify self-doubt and give you a new way to stay grounded, even when things feel uncertain.

    The Gifts—For each of the Four Attributes, a final chapter offers an expansive thought exercise to help you gain a larger perspective on your life. They are simple, have no "steps," and are available to you anytime, anywhere. They help break the spell of self-doubt and can reconnect you to a stronger sense of self.

A caveat: Self-doubt isn't one-size-fits-all. Whatever shape it takes, self-doubt is deeply personal, influenced by your past, your personality, and the challenges you're facing right now. So take what resonates, leave what doesn't, and let your approach grow as you do. Because life will change. New doubts will arise. And when they do, these tools will be right here, ready to meet the moment.

What's Ahead

First, you'll start with a twelve-question self-assessment to uncover your Doubt Profile—where you're strong, where you struggle, and how it all fits together.

Then we'll explore the deeper roots of self-doubt: how your brain is wired to protect you, and how your beliefs and internal stories often end up holding you back instead.

Next, we'll dive deep into each of the four core Attributes and how to build Big Trust one habit at a time, with lots of examples of everyday people and well-known figures—and a few of my own experiences too. I've seen this model transform careers, relationships, and entire ways of thinking, and I can't wait for you to experience it!

So, if you've ever felt stuck, second-guessed your potential, or wondered how to quiet the voice that says, *You can't do this,* you're in the right place.

Start where you are *right now*. Not where you think you *should* be. Not where someone else says you *ought* to be.

Right *here.*

Let's begin.

SELF-ASSESSMENT: YOUR SELF-DOUBT PROFILE

There's no getting around it—thinking about self-doubt touches a nerve. Your feelings (and your judgments) about your doubts can be painful, worrisome, energy-sapping, and at times overwhelming. You're probably intimately familiar with the doubts that weigh heaviest on you, the ones that hijack your confidence and trigger big emotional responses. But what you might not see as clearly are the strengths you already have and the ways you quietly keep some of your doubts in check.

That's why we begin this journey into Big Trust the same way we do with clients and students: with a simple, science-backed diagnostic. It's just twelve questions, but don't let the simplicity fool you. It cuts straight to the heart of where self-doubt holds you back most, and where your natural strengths—Acceptance, Agency, Autonomy, and Adaptability—are already working in your favor.

The point of these questions isn't only to figure out where you're thriving or struggling—it's also to bring clarity. The goal is to give you a fresh perspective on the stressful, overwhelming "blob" of your self-doubt, and turn it into something you can work with. Once you see your patterns, you can start to shift them.

And with each shift, you're building the foundation of something greater: Big Trust.

As you score your assessment, you might uncover one Attribute that feels like an Achilles' heel—that tender spot where doubt sneaks in and self-criticism hits hardest. But just as often, people discover surprising areas of strength—domains where they stay grounded and resilient, even when things don't go to plan. These insights matter.

So, grab a pen and paper, open your notes app, or just mentally gear up, because it's time to get real. Take all the time you need. Just be honest with yourself. And if you're unsure, go with your gut.

Step 1: Rate Yourself

Rate each question on a scale from 1 to 5:

1 = Strongly agree (this is exactly like me!)
2 = Agree (often like me)
3 = Neutral (sometimes like me)
4 = Disagree (not really like me)
5 = Strongly disagree (not like me at all)

Questions

1. I often feel inadequate compared to others, as if they are somehow more worthy.

Strongly agree	Agree	Neutral	Disagree	Strongly disagree
1	2	3	4	5

2. I often feel highly stressed, irritable, or anxious.

Strongly agree	Agree	Neutral	Disagree	Strongly disagree
1	2	3	4	5

3. Deep down, I worry that people overestimate my abilities.

Strongly agree	Agree	Neutral	Disagree	Strongly disagree
1	2	3	4	5

4. I find myself seeking approval from others to validate my worth and decisions.

Strongly agree	Agree	Neutral	Disagree	Strongly disagree
1	2	3	4	5

5. I often feel like other people know what they're doing, and I'm just trying to keep up.

Strongly agree	Agree	Neutral	Disagree	Strongly disagree
1	2	3	4	5

6. I often feel that I don't have control over my life.

Strongly agree	Agree	Neutral	Disagree	Strongly disagree
1	2	3	4	5

7. I set standards that are impossibly high, and then, when I fail to meet them, take it personally and become self-critical.

Strongly agree	Agree	Neutral	Disagree	Strongly disagree
1	2	3	4	5

8. I often question or second-guess my own competence or skills, unsure if I'm truly capable.

Strongly agree	Agree	Neutral	Disagree	Strongly disagree
1	2	3	4	5

9. Whether or not I succeed mostly depends on who I know, not what I do.

Strongly agree	Agree	Neutral	Disagree	Strongly disagree
1	2	3	4	5

10. I often feel like my career is out of my hands, and it leaves me feeling stuck or powerless.

Strongly agree	Agree	Neutral	Disagree	Strongly disagree
1	2	3	4	5

11. I struggle to cope with personal or work problems—they stress me out a lot.

Strongly agree	Agree	Neutral	Disagree	Strongly disagree
1	2	3	4	5

12. Even when I try to relax, my brain won't switch off—it's busy overthinking or worrying about what might go wrong.

Strongly agree	Agree	Neutral	Disagree	Strongly disagree
1	2	3	4	5

Step 2: Tally Your Score

1. Add up your responses to **questions 1, 4, and 7.**
2. Add up your responses to **questions 3, 5, and 8.**
3. Add up your responses to **questions 6, 9, and 10.**
4. Add up your responses to **questions 2, 11, and 12.**

Your Score:

Acceptance Score:	Q1 + Q4 + Q7 = ______
Agency Score:	Q3 + Q5 + Q8 = ______
Autonomy Score:	Q6 + Q9 + Q10 = ______
Adaptability Score:	Q2 + Q11 + Q12 = ______

Step 3: What Your Score Means

Each Attribute score falls into a range that tells you how self-doubt is showing up for you—whether it's running the show, just an occasional challenge to deal with, or a strength to draw on and continue to cultivate. Here's the breakdown:

Score of 15: "Superpower" Zone (Empowering). Here's where you shine. Any Attributes in this range are your secret sauce. When challenges arise, these are the parts of you that step up and say, "I've got this." Celebrate them. Use them not only to elevate yourself but also to lift up the people around you.

Score of 12–14: "Hidden Strength" Zone (Beneficial). These are your quiet wins. In these areas, you're generally confident and self-assured, and while self-doubt might show up for an unwanted cameo now and then, it doesn't hang around long enough to do real damage. They're a solid foundation you can build on to tackle other areas of growth.

Score of 9–11: "So-So" Zone (Moderate). This is the "It's not ruining my life, but it's definitely not helping" territory. With a little focus, you could easily turn these into strengths that actually work for you, not against you.

Score of 7–8: "Hindrance" Zone (Detrimental). A score in this range means that the Attribute is eroding your confidence and setting up roadblocks, often without you even realizing it. This is a call to action for some nurturing and support to turn these detrimental patterns around before they do any more damage.

Score of 6 or Less: "Red Alert" Zone (Destructive). This is where self-doubt is taking up so much space in your life that your whole sense of self, your ability to make choices and decisions, your relationships, and your emotional state are affected. If any Attribute score lands here, it's a blaring alarm bell saying, "Pay attention!" This is your priority area, the place where intentional effort can help break these negative patterns.

A Short Primer on the Four Attributes

Self-doubt follows specific patterns directly tied to how you see yourself. In other words, it pulls on four specific levers that determine your sense of self-trust. These four Attributes shape how you show up—in everyday moments and in those big, defining ones. They influence whether you lean in with confidence, operate in your peak state, and live with intention, or whether self-doubt pulls you back into playing small. You'll come to understand each of these Attributes in much more depth in later chapters. For now, a simple understanding is all you need to get a clearer picture of your own Doubt Profile.

ACCEPTANCE answers the question *Am I enough?* As a strength, this Attribute represents your internal sense of worthiness and acceptance of your strengths, your quirks, and your flaws, and knowing that your worth isn't tied to external approval.

When Acceptance is at its lowest, what takes its place is *self-rejection,* where every imperfection feels like undeniable evidence that you're unworthy. You constantly question your value, nitpick yourself to pieces, bend over backward for other people's approval, chase impossible perfection, and freeze at the mere thought of failure.

If you wrestle with Acceptance, deep down you may believe:

I'll never be enough unless I'm perfect.

I can't accept this part of myself unless others do.

I have to change who I am to feel valued.

AGENCY answers the question *Can I handle this?* This reflects your belief in your own readiness—your trust in your ability to take action, solve problems, and follow through. Agency is what fuels your momentum, helping you rise to a challenge with confidence and persistence.

When Agency is at its lowest, what you feel is *inefficacy,* where every obstacle feels like proof of your inability to create a successful outcome and that you just don't have what it takes. You doubt your skills, feel like a fraud, compare yourself to everyone who seems to have their act together, and end up stuck in a cycle of hesitation that keeps you from moving forward.

If you struggle with Agency, deep down you may believe:

Everyone else seems to have it all figured out—they're so much further ahead.

Seeing others succeed where I struggle just proves I don't have what it takes.

If I admit I don't know something, everyone will see me as a fraud.

AUTONOMY answers the question *Do my choices make a difference?* This Attribute relates to taking ownership—owning your actions, your choices, and the consequences that come with them. It's the belief that your decisions shape your future and that you have the ability to influence the direction of your life.

When Autonomy is at its lowest, you experience *resignation,* where life feels like a series of events happening *to* you, not because of you. You end up stuck in a cycle of helplessness. You play it safe, shrink into your comfort zone, avoid risks, and convince yourself that the world is out to get you.

If you struggle with a sense of Autonomy, deep down you may believe:

I don't really have much say in whether I succeed or fail.

Life is a lottery—and I always seem to lose.

Life's not fair. I can't change anything, so what's the point of trying?

ADAPTABILITY answers the question *Can I manage my emotions?* This Attribute is your emotional anchor. It reflects your ability to stay grounded when life gets messy, regulating your emotions instead of being ruled by them.

When Adaptability is at its lowest, you descend into *overwhelm,* where the smallest hiccup throws you off course. Stress piles up, anxiety takes over, and every little setback feels like the end of the world. You lose perspective and can't think clearly. Instead of moving through your emotions, you get stuck in them.

If you struggle with Adaptability, deep down you may believe:

My emotions control me more than I control them.

Feeling upset is a trap with no exit; once the emotions flood in, logic checks out.

I'm just not cut out for dealing with pressure.

Your Doubt Profile is a mirror reflecting the exact areas *where* self-doubt takes hold and *how* to rebuild self-trust in its place. It shows you where to focus, which Attribute is most impacted, and which habit to develop to strengthen it. Because while Attributes reveal how doubt shows up in your life, it's the habits you practice that determine whether it stays or fades.

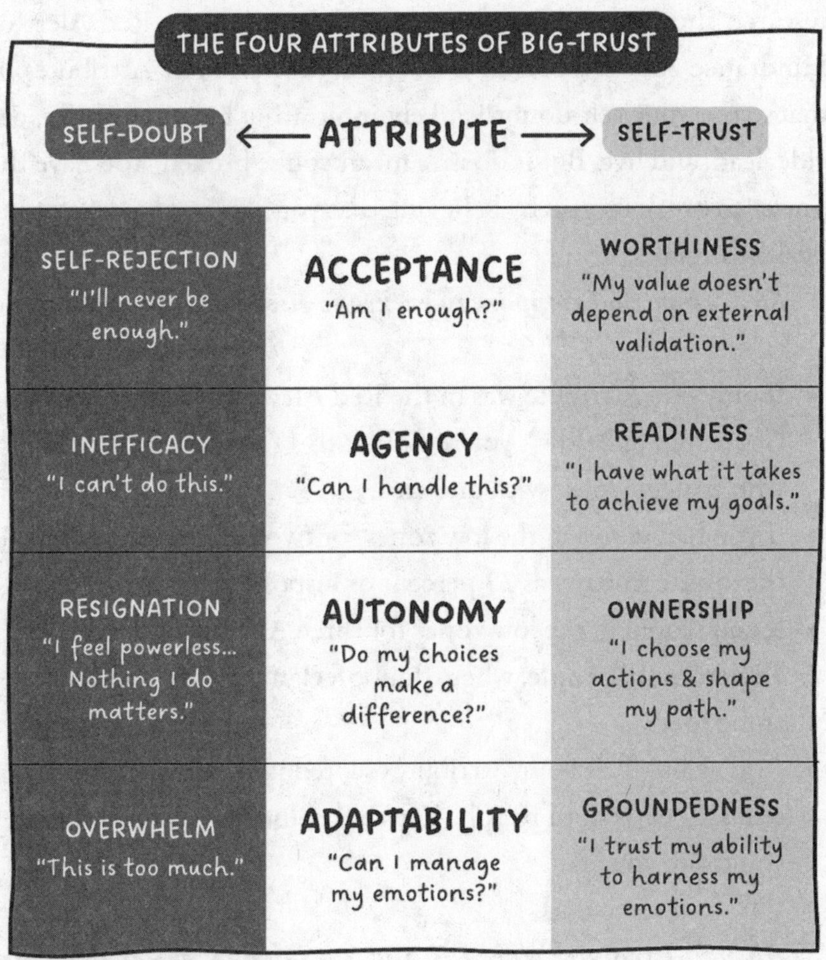

THE FOUR ATTRIBUTES OF BIG-TRUST

SELF-DOUBT	← ATTRIBUTE →	SELF-TRUST
SELF-REJECTION "I'll never be enough."	**ACCEPTANCE** "Am I enough?"	WORTHINESS "My value doesn't depend on external validation."
INEFFICACY "I can't do this."	**AGENCY** "Can I handle this?"	READINESS "I have what it takes to achieve my goals."
RESIGNATION "I feel powerless... Nothing I do matters."	**AUTONOMY** "Do my choices make a difference?"	OWNERSHIP "I choose my actions & shape my path."
OVERWHELM "This is too much."	**ADAPTABILITY** "Can I manage my emotions?"	GROUNDEDNESS "I trust my ability to harness my emotions."

What Do My Scores Mean for Me?

Your score is not good or bad, pass or fail. It's a new way to understand your strengths and vulnerabilities. Here are answers to some of the most common questions we hear about what these scores mean:

I scored low across all Four Attributes. What does that say about me? First of all, take a breath. You're not alone. In fact, our research shows that 20 percent of people are in the Red Alert or Hindrance zones (scores of 8 or below) for all Four Attributes. If that's you, your self-doubt is likely infiltrating how you think, decide, lead, and live. But it doesn't mean you're broken. You have the power to break its grip by believing that you can and by developing healthier habits.

So . . . how do I compare to everyone else? Let's break it down:[1]

- If only one Attribute was in the Red Alert or Hindrance zones (score of 8 or below), you're similar to 17 percent of people who struggle most with one area.
- If you had scores in the low zones for two Attributes, you're in the same company as 21 percent of respondents.
- If you scored in the low zones for three Attributes, you're like 18 percent of people, where doubt feels a little more deeply embedded.
- And again, if all Four Attributes scored in those low zones, that's 20 percent of people. You're absolutely not the outlier you think you are.

All my Attributes scored in the So-So zone. What does that mean? This suggests you're not fully anchored in any one area of self-trust just yet, which can make your self-doubt feel inconsistent. It might flare up, depending on who you're around, where you are, or even how you're feeling that day. You're right on the edge. With a few intentional shifts, you can start to strengthen your self-trust so it's more stable, no matter the situation.

All my Attributes scored in the Hidden Strength or Superpower zones. What does that mean? That's a great place to be. With strong Attributes, you likely have a rock-solid foundation of self-trust—what we call Big Trust. This doesn't mean you never doubt yourself (you're human, not a robot), but it does mean you've built habits that help you understand and use your doubts to help you, and when they're getting in your way, you're able to move through them relatively quickly. Keep reinforcing those habits—they're your edge. And if you're in a leadership, parenting, or mentoring role, think about how you can help others strengthen these Attributes in themselves too. Self-trust is contagious when modeled well.

What if some of my Attributes scored high, but others scored low? That's actually very common, and it's something you can use to your advantage. You can lean on your stronger habits to help support the ones where you scored lower. Say you have strong Agency (you take action even when it's hard) but low Acceptance (you struggle with self-worth). You can use your action-taking habits to build evidence of your value. Each Attribute supports the others, so when you strengthen one, you often lift the rest with it.

The following bar charts show how people score on each of the Four Attributes. Each bar is split into three zones: Red Alert or Hindrance (left), So-So (middle), and Hidden Strength or Superpower (right).

• • • •

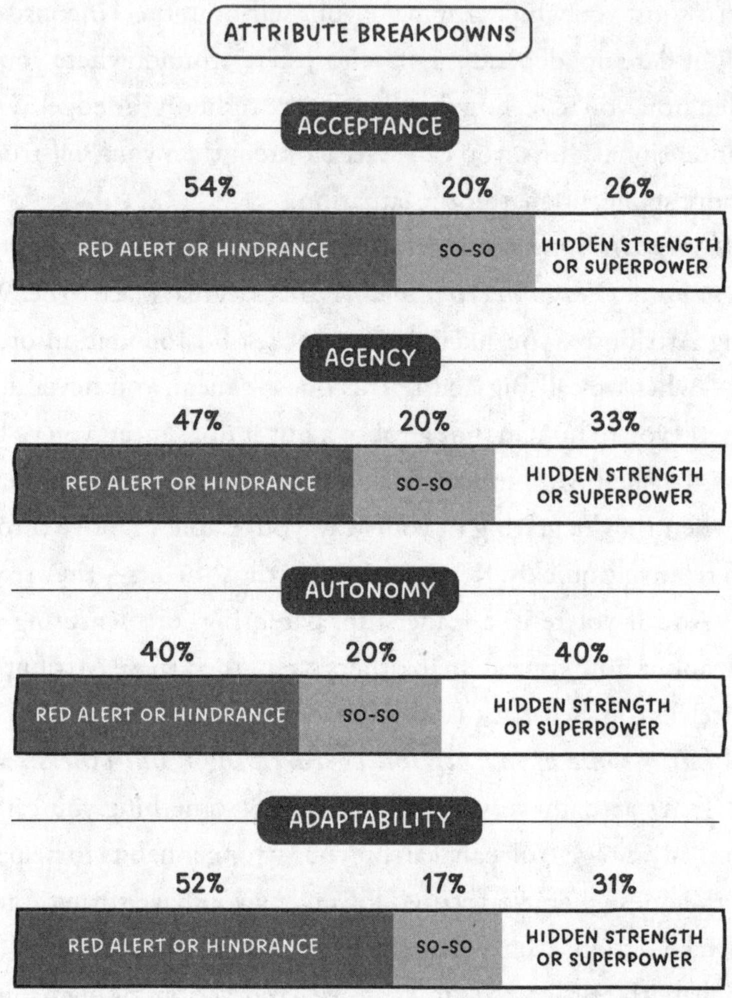

These Attributes lay the groundwork for self-trust. They quietly shape how you show up—in everyday moments and in those big, defining ones. They influence whether you lean in with confidence, operate in your peak state, and live with intention, or whether self-doubt pulls you back into playing small.

You don't need to let doubt define you and run your life. You can use it as a cue to learn, grow, and ask for help. You can interrupt those exhausting spirals of doubt—but first, we need to lay the foundation.

In chapters 1 through 3, I'll walk you through how your brain's doubt wiring works, because once you understand *why* self-doubt takes hold, you'll be far more equipped to overcome it. I can't wait to get you working on the solution, but this background will make the steps ahead land deeper, hit harder, and stick longer.

Then, starting in chapter 4, we'll take a closer look at each of the Four Attributes—what they are, how doubt undermines them, and how to strengthen them—so you can identify which areas need your attention most. Because the more you strengthen these Attributes, the closer you move toward Big Trust.

THE SELF-DOUBTING MIND

CHAPTER 1

Who's Running the Show?

Hardwired for Self-Doubt

Self-doubt is baked into the human experience, like taxes or your phone dying right when you need it most. The problem isn't that self-doubt exists. The real problem is *what you do with it.*

One of the many truths I've learned through studying doubt is this: Self-doubts are not immovable facts. They are the products of your brain's wiring, your formative experiences, and your internalized stories about your identity. In other words, you are not your self-doubts, and your doubts don't define you.

Some of you may need to read that again. Slowly.

You are not your self-doubts. Not even close.

Self-doubt isn't about what you can or can't do. It's about what you *think* you can't do. Self-doubt shapes your reality, but it's your *thinking* that shapes self-doubt. Your thinking gives doubt its power and wears you down. To change your doubt, change your thoughts.

Now, I know that's easier to put into words than to put into practice. It's your brain's fault, actually. Through my work, I've discovered that we can't escape self-doubt for three primary reasons that are hardwired into every one of our brains. When you understand how your brain is designed for self-doubt—the neuroscience and psychology behind it—you can develop a new relationship with it. In doing so, you will deepen your trust in yourself and loosen the grip of self-doubt.

Your Brain Needs to Manage Itself

Marco was a software engineer at a billion-dollar big-tech company, but he had a dream to launch his own ambitious start-up. He had it all lined up—the research, a solid business plan, and even investors. But there was a problem—he couldn't take the leap. The pressure of doing it all on his own had him paralyzed. Instead of taking action, he buried himself in an endless cycle of second-guessing every decision and diving into more research. His inability to move forward reinforced the deep-seated thought he'd had all along: *Maybe I can't do this after all.*

In the throes of that anguish, Marco visited me—and we visited his boardroom.

Here's a helpful metaphor I often use with my clients. It's based on the work of Professor Julius Kuhl, a leading authority in human motivation, who has dedicated his career to uncovering what drives people to take action—and what holds them back.[1] So, picture your brain as a company boardroom.

As in any boardroom, each member has a role to play.

- The CEO (located in your ventromedial prefrontal cortex) thinks about the future and goals, and asks, "What's most important right now? What's the vision?"

- The Strategist (run by the left prefrontal cortex) helps make the game plan to achieve those goals and asks, "How do we make this happen? What comes next?"
- The Automations Lead (managed by the basal ganglia and cerebellum) helps make everything run more efficiently and asks, "What routines can we set up to make this easier?"
- The Risk Analyst (in the inferotemporal cortex and the limbic system) is there to protect you and asks, "What could go wrong? How could we address it?"

When all four members of the board are in sync, your brain operates like a dream team. Goals are clear, plans are actionable, tasks are automated, and risks are handled. Like an athlete in a flow state, you feel your best and your brain works best when its functions are coordinated.

The problem is that your brain has multiple complex systems all vying for attention and priority. When self-doubt takes hold in any system, that flow dies. Departments stop communicating, priorities get muddled, and the operation grinds to a halt. Your brain has an office-wide meltdown.

And that's exactly where Marco found himself. The once-clear vision of his mind's CEO had become cloudy. No direction forward felt right or secure. Without clear direction, his Strategist had plunged into a spiral of overthinking, resulting in analysis paralysis and halted progress. Without plans to enact, his Automations Lead was just waiting around, twiddling its thumbs. And witnessing the disarray across the boardroom, his Risk Analyst thought, *What the heck are we doing here?!*

Marco and I worked together to sort out the chaos of his board members and pinpoint where self-doubt had hijacked his systems.

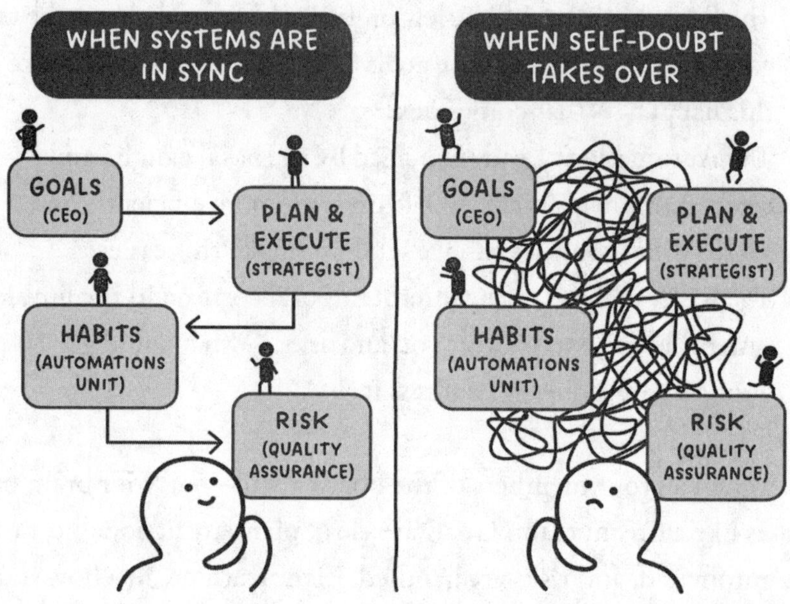

We started at the top with his CEO, which only had a vague notion of "launching soon," and the wheel-spinning was creating dread and even more self-doubt. We got crystal clear and specified a new goal: Secure the first paid pilot within six months. With a clear goal, his Strategist could focus on relevant, achievable next steps: identifying key logistics partners and preparing a targeted pitch for them. With plans in place, his Automations Lead reestablished a structure for their work going forward, including weekly progress reviews. Feeling more secure in the goals, strategy, and plans, his Risk Analyst stopped overreacting. Instead of spiraling, Marco practiced pausing and asked himself, "Is this a real problem? Or is this just fear talking?"

Once his mental boardroom became more coordinated, Marco felt like he was back in control. He started making progress without spinning his wheels and doubting every action. He could trust himself and could believe in his big idea. Plus, once he formally launched, he would be in a much better position for the future. Ultimately, he stopped standing in his own way.

He told the boardroom of his brain who was truly in charge.

PRACTICE: Sync Your Boardroom

Is your boardroom on overdrive? Has it shut down from burnout? Is it in general disarray? When you recognize that your brain's boardroom is out of sync, you will gain the power to step back and recalibrate. Here's what to watch for, along with small actions to help you reset and get back on track:

- **For the CEO:** Are you feeling scattered, like you're juggling too many priorities and none of them are clear?
  - One action: Write down your top priority for the next week. Just one. Then make sure every task aligns with it. If it doesn't, it's not a priority.
- **For the Strategist:** Are you stuck in analysis paralysis, endlessly planning and weighing options without taking action?
  - One action: Break your project into three simple steps. Just three. Now commit to completing the first one today (or this week, if you need to pace yourself).
- **For the Automations Lead:** Are you constantly feeling drained because tasks that should be automatic require extra effort and decision-making?
  - One action: Identify one routine task you do regularly—like checking emails or prepping for meetings—and create a simple checklist to streamline it, such as organizing your emails into categories (Follow Up, Read Later, Pending) or setting a specific time to prep for meetings.

- **For the Risk Analyst:** Are you constantly imagining worst-case scenarios or feeling like small mistakes will lead to disaster? (We'll dig deeper into this later in the chapter.)
  - One action: List out the worst-case scenario and then write down what you would do if it happened. This helps shrink the fear to its real size.

Your Brain Craves Certainty

Many of us go through life in a state of constant high alert, regardless of what's actually happening. It can feel like our boardroom is a mess, making it hard to function at our best. Why do we enter this state in the first place? Often, it's because our brains desire certainty. But this innate drive clashes with the reality of modern life.

Here's how I like to look at it. When a tomato plant gets attacked—say, by a caterpillar munching on its leaves—it doesn't just accept defeat. The plant kicks into survival mode by triggering a highly efficient defense system. It sends out distress signals, toughens up its leaves, churns out proteins to make itself harder to chew, and fortifies its cell walls. It's the botanical version of "lock and load." But once the caterpillar is gone, the tomato plant doesn't stay on high alert. It relaxes. It shifts its energy back to growth and repair. That's why the plant thrives. It knows when to protect itself and, just as crucially, when to stand down.

You also have a built-in defense system. It's called self-doubt, or anxiety, or overthinking. Basically, your brain's Risk Analyst raises the alarm when it thinks something's off. The problem? Unlike the tomato plant, your Risk Analyst is *terrible* at distinguishing real threats from imagined ones. Danger or not, it's a threat-detection machine on steroids. And it's even worse at letting you know when a potential threat has passed. You end up on high alert over a social media post that got zero likes, a typo in an email, a sideways glance, or, worse, something that hasn't even happened.

There's an evolutionary reason for this. This threat-detection function was a lifesaver for our ancestors. Spotting a predator in the bushes or a storm on the horizon meant the difference between seeing another sunrise or becoming part of the food chain. Fear had a clear purpose. *Spot the threat. Act fast. Survive.* Over time, our brains became skilled at spotting patterns and predicting what happens next. They became so good, in fact, that they almost developed an "addiction" to certainty. Knowing what to expect was not only comforting, it was a massive survival advantage. Certainty kept us safe in a world where not knowing could get you killed.

Today, the world is infinitely more complicated, but our brains still run the same old survival software. Only now, they obsess over whether your boss liked your presentation or if that awkward comment during the client meeting lost you their business. The threats aren't life-and-death anymore, but your brain doesn't care. It still treats uncertainty like danger because that's what it is wired to do.

Evolutionary scientists call our modern world a "delayed-return environment."[2] In the past, you'd hunt, gather, or run, and the reward—or consequence—was immediate. But now? You work hard for a promotion that *might* come next year. You send an email and wait for a reply that *might* never come. Your efforts are tied to long-term payoffs: building a career over years, maintaining relationships, saving for retirement for your future self. The feedback loop is gone, replaced by a gaping hole of uncertainty.

Your brain hates that.

In fact, your brain hates uncertainty so much that it will fabricate the truth to make itself feel better. It fills in the blanks to make your life feel more predictable. Whether its answers are accurate or not, your brain autocompletes the story. Psychologists call this "confabulation," from the Latin *fabula* for "story." But let's just call it what it is: your brain's often faulty autofill.

Your brain automatically invents explanations that seem logical. For example, you may tell yourself, "I'm just not a people person," so you skip networking events. In reality, you're just nervous about coming across as awkward, and your brain is trying to protect you from feeling judged. Or, you may say, "I'm fine where I am career-wise." But the reality is you don't want to compete with a colleague for a coveted assignment because you're petrified that if you get it you'll damage that work relationship.

These confabulated mental stories *feel* logical and comforting because they provide certainty. But peel back the layers and you'll find they're powered by fear, not truth. As Professor of Law Ian Weinstein put it: "Don't believe everything you think."[3] Because your brain is wired to prioritize threat detection over accuracy, these mental stories skew negative. This is why your brain often defaults to worst-case scenarios:

They didn't reply because they hate me.

That meeting went badly. I'm getting fired.

No one liked my post. I'm irrelevant.

None of these are facts. They're just your brain's attempt to regain control in an uncertain situation, even if that "control" comes wrapped in anxiety and nonsense.

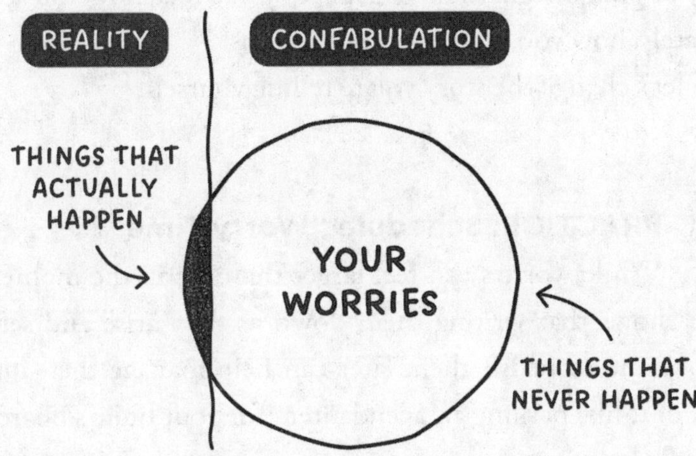

Instead of calmly evaluating situations, your brain leaps straight into disaster mode to create the *illusion* of control. This mental habit doesn't just waste energy, it also turns into chronic hypervigilance.[4] Worry becomes a way of life. Your brain convinces itself that if you just worry hard enough, you can prevent bad things from happening. But worry doesn't solve problems—it just catastrophizes about ones that probably don't exist.

And these confabulated stories come with a cost. Harvard researchers Matthew Killingsworth and Daniel Gilbert conducted an experiment that discovered something staggering: Your thoughts have a greater influence on your happiness than your actual circumstances.[5]

Let that sink in. You could be lounging on a beach in paradise, but if your brain is spiraling into worry—replaying fears, doubting yourself—you'll miss the sun, the breeze, and the sound of the waves entirely. Your thoughts don't just *color* your experience; they *define* it. This means that the stories you tell yourself—whether rooted in reality or pure imagination—shape how you feel, how you act, and, ultimately, how you experience life.

So let's change the story you're telling yourself.

PRACTICE: Schedule "Worry Time"

While worries can feel larger than life in the moment, research shows that writing them down as they arise and setting a specific time to revisit them later can help manage their impact.[6] Think of it like parking an agenda item for your brain's boardroom to handle later.

Step 1: Create a "Worry Time" Zone

Pick a specific time and place each day to focus on your worries. Maybe it's 5:30 p.m. on your balcony, or 7 p.m. on your couch with a cup of tea.

Step 2: Park Your Worries

When an anxious thought starts gnawing at you, don't spiral. Write it down. Scribble it in a notebook or add it to a worry list on your phone. Then tell yourself, "Not now. I'll deal with this at Worry Time."

Step 3: Dive Into Worry Time

When it's Worry O'Clock, pull out your list. Give yourself up to thirty minutes to dig into your concerns. Let yourself feel whatever comes up—no judgment. Add any new worries if needed. When your time's up, that's it. Close the notebook, shut the app, and walk away. No more worrying till your next Worry Time.

Step 4: Decide What's Next

Now that you've aired out your mental clutter, ask yourself:

- Are these worries real, or am I catastrophizing?
- Can I take any meaningful action right now to address this?

If the answer is yes, make a plan. Commit to one small action. This almost gives your brain an immediate return on the effort. If the answer is no, shift your focus to something more productive.

Step 5: Periodically Review Your Worry List

At the end of a week, review your worry list. Notice any patterns? How many worries turned into real problems? This practice helps you spot the exaggerations in your thoughts and keeps worry from taking over. You're telling your brain's Risk Analyst, "Okay, you can have your time, but the rest of the day belongs to me."

Your Brain Filters Reality

In my early banking days, I was briefly an executive assistant to a general manager who led a sales team of more than two thousand people. Part of my job was protecting his focus. I was his gatekeeper. I filtered his email, sorted the urgent from the important, and made sure only the most critical matters landed on his desk. It was no small task. What struck me most was how essential this role as gatekeeper was. Without it, his priorities would have been buried under a mountain of distractions.

Your brain's boardroom has a Gatekeeper too, the ultimate controller of your attention. In addition to the four systems that make up your brain's management system, a neural network acts as your ultimate filter.[7] This network determines what grabs your focus and what fades into the background. It feeds that information to your CEO, ensuring you prioritize what seems most important.

Researchers estimate that your brain is bombarded with the equivalent of thirty-four gigabytes of data every day (that's a lot!).[8] Without your Gatekeeper, you'd drown in the noise, overwhelmed by every car horn, random thought, or phone notification. But for all the good your Gatekeeper does, it has a major fault: It's not objective. It doesn't prioritize based on logic alone. No, it takes its cues from you—specifically, from what you're *already* focused on.

I'm sure you can see where this is going.

If you're stuck in self-doubt, fear, or worry, guess what? Your Gatekeeper tunes into that frequency. It starts filtering reality through the lens of your insecurities, amplifying every little detail that reinforces those doubts. It puts a spotlight on the "evidence" of your shortcomings. Suddenly, an unanswered text feels like rejection, every typo feels like failure, and every challenge feels insurmountable. Your brain becomes a self-doubt amplifier. Over time, this warped filter doesn't just mess with your thoughts; it reshapes how you see yourself and how you show up in the world. This is the power, and the danger, of your focus.

Taylor offers a good example. She was a senior leader at a telecom company, but she struggled every day with her perceived shortcomings. She was a steady, quiet high achiever—reliable, detail-oriented, the kind of person everyone knew they could count on. A few years into her role, a mentor gave her what was *supposed* to be helpful feedback: "You need to be more assertive if you want to move up."

But Taylor didn't hear, "You've got potential." She heard, "You're not leadership material." She didn't see it as guidance to help her grow; she saw it as a judgment about *who* she was. Just like that, her brain's Gatekeeper locked onto that belief. From then on, she filtered everything through that lens:

A skipped invite to a casual meeting? *I'm not wanted.*

A manager suggesting her presentations could be more concise? *I'm not a clear communicator.*

A colleague interrupted her in a meeting? *My ideas aren't valuable.*

It didn't stop at work either. When a friend replied with a short, rushed text, Taylor's brain didn't think, "Oh, they're probably busy." Nope. It whispered, "You're not a priority. See? You're forgettable." Her Gatekeeper kept collecting "evidence" *everywhere* to support the narrative. And the more it found, the more deeply Taylor believed it.

Taylor's story isn't unique. It's just how our brains operate. We do this with so many things—health trends, pop culture, politics, you name it. It's a cognitive distortion you know as *confirmation bias*. Your brain's Gatekeeper is a master at zooming in on information that supports what you already believe and quietly filtering out anything that doesn't.

Here's the deeper problem: When those beliefs are rooted in self-doubt, your Gatekeeper turns fleeting moments into lasting evidence, transforms opinions into facts, and magnifies fear until it feels like truth. So if your Gatekeeper keeps lying to you, how can you discover the truth?

Your Gatekeeper has one job: Filter all information before it gets to your boardroom. The good news is that your Gatekeeper can be retrained. The key is teaching it what to notice.

When Taylor and I started working together, she believed she constantly had to prove herself and that every little delay in recognition meant she wasn't valued. Her boardroom was drowning in negativity because her Gatekeeper was feeding it a steady stream of self-doubt and worst-case scenarios.

I told her, "We need to invite two more professionals into your boardroom: the Fact Checker and the Confidence Consultant. Consider them like outside consultants who'll help your Gatekeeper do their job better."

Here's how that looked for Taylor:

Call Out the Confabulation. She wrote down every worry and insecurity she had. Then we challenged each one by inviting her internal Fact Checker to the boardroom. This is an internal voice of reason who's unafraid to verify the truth of your situation. When Taylor told herself, "No one wants me here," we pushed back like a defense attorney: "Where's your proof? Where's the evidence? And what's the evidence that says otherwise?" This simple fact-checking exercise opened her eyes to the critical realization that most of her doubts weren't grounded in reality. They were confabulations, stories her brain had autofilled from assumptions, past experiences, worries, and fears.

Shift the Focus. Then Taylor actively shifted her Gatekeeper's focus by inviting her Confidence Consultant into her boardroom. To fulfill this role, I asked Taylor to envision a trusted mentor, friend, or colleague who always sees her potential and strengths. In moments of doubt, she could imagine this friend sitting across the table from her, giving her straightforward but encouraging advice, telling her what they thought about her ability to handle the situation.

Taylor embraced the practice. When silence used to feel like rejection in meetings, she heard her Confidence Consultant lean in and say, "Silence isn't rejection. It's reflection. Give them space to process." Instead of spiraling into "They hate my proposal," she heard that steady voice guiding her to ask, "How can I make this clearer?" Even receiving negative feedback, which once felt like a personal attack, transformed too. She imagined a mentor's steady voice saying, "This isn't about your worth. See this as an opportunity for growth." She stopped seeing feedback as a verdict on her value and started viewing it as a tool for improvement.

This shift impacted Taylor's personal life too. She finally began cutting herself some slack. If a friend got distracted mid-conversation, she didn't jump to "I'm boring. They don't care." Instead, her Fact Checker kicked in: "What's the evidence for that?" It reminded her, "People have their own stuff going on. They're probably just busy. Not everything's about you."

As Taylor redirected her Gatekeeper with the perspective of her Fact Checker and Confidence Consultant, she noticed a change within. She carried herself with more confidence and doubted herself less. She told me, "For the first time in my life, I feel like I'm on *my own side*."

PRACTICE: Invite Your Internal Allies

When self-doubt hits, invite your Fact Checker and Confidence Consultant into your boardroom to balance the chaos. These are your mental allies.

Step 1: Call Out Your Central Confabulation with the Fact Checker

A. Write down your assumption (e.g., "If I make a mistake, everyone will think I'm incompetent," or "I'm always the last person chosen for teams, so it must mean no one wants me here").

B. Run your assumption through these three questions:

- What evidence do I have that this is true?
- What evidence do I have that this is *not* true?
- What's a more balanced way to look at this?

Step 2: Shift Your Focus with Your Confidence Consultant

A. Imagine a trusted mentor, friend, or colleague sitting across the table from you, someone who always sees your potential and strengths and lovingly cuts through the noise of your life. Ask yourself, "What would they say about my ability to handle this?" Then hear their voice reminding you of your competence and resilience.

All of us are wired for self-doubt—but what's wired can be *rewired*. When you resync your boardroom, you gain clarity and experience flow. When you schedule "Worry Time," you free up mental bandwidth for more meaningful thoughts and you avoid becoming consumed by confabulations. And when you invite "consultants" to keep your Gatekeeper focused on your best path forward, you focus less on your weaknesses and more on your strengths, your wins, and your potential.

When you start recognizing how your brain works and how it primes you for self-doubt, you stop fighting yourself and start showing up for yourself. And you finally feel like you're on your own side.

CHAPTER 2

Should I Believe Everything I Think?

Hooked by Your Self-Image

In the late 1970s, Dartmouth psychology professor Robert Kleck set out to test a simple but powerful idea: *Do our expectations shape the way we experience the world?* He conducted one particularly fascinating experiment that revealed just how deeply our own beliefs can warp our reality—and this is especially revealing when it comes to self-doubt.[1]

He took a group of participants and applied a large facial scar to their right cheek, stretching from ear to mouth. Each person was shown their reflection in a hand mirror, confirming that they now had a visible disfigurement. Then, they were sent into conversations with strangers. Meanwhile, a control group had no scars applied and also interacted with strangers.

After the conversations, both groups were asked by the researcher to describe how they felt they'd been treated. Those with the scar overwhelmingly reported that the strangers had acted differently. They were more tense, less friendly, even somewhat distant.

The scarred individuals were convinced that the scar had changed how others saw them.

But it hadn't.

Because the scar never existed.

Just before the scarred participants left, the experimenter pretended to apply some moisturizer to "set" the scar. Instead, they secretly removed it. These participants walked into their interactions *believing* they had a noticeable scar when, in reality, their faces were untouched. Neutral observers watching the same interactions on video saw no differences in the strangers' behavior between the groups. The change wasn't in how others acted—it was in how the participants *perceived* it.

This is what psychologists call *expectation bias.*[2] The study participants believed they looked differently, and their brain projected that insecurity onto reality. They saw rejection, judgment, or discomfort—not because it was there, but because they *expected* it to be.

Your beliefs shape your reality. *Literally*.[3] Your thoughts have *that* much power.

In *Psycho-Cybernetics,* surgeon-turned-writer Dr. Maxwell Maltz explained: "Whether we realize it or not, each of us carries around with us a mental blueprint or image of ourselves . . . our own conception of the 'sort of person I am.'"[4] And this blueprint influences how we navigate life. Maltz had seen this firsthand with his plastic surgery practice. Patients would come to him to fix a "flaw" in their appearance that they were deeply insecure about. Their thinking was pretty straightforward: *If I fix this flaw, I'll feel more confident. If I look better, I'll be more successful.* But that's not how it played out.

After surgery, even with the physical change, many patients still acted, felt, and thought as if the flaw were there. Why? Because their brains never updated their blueprint. Their beliefs about themselves hadn't changed, so their self-doubts lingered.

And here's where it hits home. You could have a trail of achievements behind you, landing promotions left and right, winning awards, securing funding for your start-up, starting a family, but if deep down you believe you're *inadequate* or *unworthy,* none of those accomplishments feel real. You'll fixate on what's missing. You'll avoid opportunities, back out of challenges, ruminate, and turn your strengths into liabilities. Like Kleck's research subjects and Maltz's patients, your actual flaws aren't holding you back—it's your *belief* in them.

All the data backs it up. Who you *believe* you are is the blueprint for your entire life.

Sticky Stories

When I first met Rashida, she introduced herself with a disclaimer: "I'm a little intense." She said it with a grimace, as if the label left a bad taste in her mouth.

I replied, "Good to know. What else should I know about you?"

She told me she was a mother, a recent pickleball enthusiast, and a leader in risk and compliance at a Fortune 500 company. I thought maybe such a role demanded intensity, but I still asked, "Where does that 'intense' label come from?"

She didn't have to think long. "When I left my last company, my boss made an offhand comment. He said, 'You can be a bit intense, Rashida, but we'll sure miss you.'"

That comment stuck. Hard. But it wasn't the first time Rashida had heard it: "My parents are Egyptian and I'm the youngest of eight kids. My brothers and sisters would call me 'too much' or 'over the top' all throughout my childhood." Over the years, those words had lingered, shaping how she saw herself. Even as a successful leader, she felt that "intense" label hovered over her. Her boss's comment had brought it all back.

Fast-forward six months and Rashida was stepping into a senior lead compliance role—a big step up. She felt compelled to introduce herself in a way that preemptively softened any judgment. She'd say, "I can be intense at times," as though she needed to apologize for being herself.

As though her invisible scar defined her.

Once a label sticks, it feels real. Take "I'm bad at public speaking." It's not because you're biologically incapable of forming coherent sentences in front of people, but maybe on one occasion you stumbled over your words, felt embarrassed, and that experience turned into an identity. You kept replaying the story until it became the headline of your personal narrative.

Why do we do this to ourselves?

Because your brain loves shortcuts. They're efficient. You label things to make sense of them faster: *hot/cold, safe/dangerous, success/failure.* But you don't just label the world around you. You slap those labels on *yourself* too. Most of your labels are lazy shortcuts your brain took back when you didn't know better. Others are ones you've picked up along the way and casually stuck to yourself without a second thought. "I'm bad at this," "I'm such a procrastinator," "I'm a terrible communicator."

Every time you say, "I am . . ." you're sticking a label on yourself—one your brain interprets as permanent and unchangeable. But that's the language of a fixed mindset,[5] one that convinces you that your abilities are set in stone, that who you are today is all you'll ever be.

But that's not true.

Peeling Off Your Self-Limiting Labels

Like burrs on a hike, once labels latch onto you, they're a pain to get rid of. But they're not permanent. They're not tattooed on your soul, no matter how much they might feel that way. Labels like "incompetent," "flawed," or "screwup" can be peeled off, examined, challenged, and rewritten.

Think back to the labels you might've been given as a child. Maybe someone called you "stubborn." But was it really stubbornness, or was it determination, which has a whole different energy? Maybe a teacher labeled you as "shy," but what if you were actually observant and thoughtful? Or when your ex-partner called you "indecisive," was it actually that you value careful deliberation?

Whatever labels you've been carrying, you can peel them off. Challenge them. Rewrite them. Replace "I'm not confident" with "I'm learning to be more confident." Swap "I'm not a leader" for "I'm figuring out what leadership looks like for me." It might sound like wordplay, but it's not. These shifts are tiny rebellions. They tell your brain (and anyone listening) that you're *not* stuck. You're evolving.

Now, I'm not claiming you can radically change how you see yourself just by swapping a label. Peeling back deeply rooted labels takes time and effort—and sometimes therapy. What I *do* know is that every time you loosen the stickiness of a label, you create space—space to reclaim who you really are, and who you want to be.

For example, I asked Rashida to reframe how she introduced

herself. Instead of "intense," I suggested "passionate." Her face lit up. "You know, for so long, I clung to that 'intense' tag, and it always felt negative. It made me doubt myself, and I almost didn't take this job because of it. But you're right. It's not intensity; it's pure passion! I'm deeply committed to my work, and I care about doing the right thing."

That shift in wording wasn't even about the word itself. It was the start of shifting how she saw herself. Suddenly, she wasn't "too much." She was driven. Committed. Exactly the kind of person you'd want leading compliance—someone who doesn't brush off the details because she cares.

I love Meg's story too. Ever since she was young, Meg was anxious. When she was a child, her grandfather joked that her curly hair was from worrying too much. Her dad called her "Moody Meg," as if her anxiety was her defining feature. For years, she wore that label like a name tag: *Hi, I'm Anxious.*

But as Meg grew older and leaned into her writing, she had a powerful realization. The same imagination that fueled her anxiety also fueled her creativity. She didn't get rid of the label; she *reclaimed* it. She said, "I'm done with condemning my anxiety and saying it should go away . . . It's a big part of why I am who I am and what I've been able to accomplish and give to the world."[6]

And what did she give to the world? Stories that have helped millions of people understand their own emotions. Meg is Meg LeFauve, one of the screenwriters for Pixar's *Inside Out* and *Inside Out 2,* beloved animated movies about identity, self-acceptance, and learning to embrace all the parts of ourselves. Meg reclaimed Moody Meg as a creative trailblazer teaching children and adults about the world of emotions.

Meg and Rashida didn't "fix" themselves. They didn't need fixing. They reclaimed their labels. They realized the words weren't the problem—it was the meaning they'd attached to them.

PRACTICE: Reclaim Your Labels

First, take a moment to think about the labels you've been carrying around.

- What labels have you internalized?
- Where did they come from? Who put them there?
- Are they helping you or holding you back?
- How do these labels influence your thoughts, actions, decisions, relationships, and even goals?

Most of us don't realize we're wearing these labels until we stop and look. And once you spot them, you get to decide: *Do I really want to keep this?*

Second, the moment you catch yourself in the act of internal "name-calling," consciously switch to a statement that affirms a strength or your capacity to grow. You'll find that over time, these new ways to describe yourself can reshape your self-image and snowball into massive changes in how you see yourself.

Here's how you can start: Pick one negative "I am" label that's shaping how you see yourself. Then flip it into an affirming, growth-oriented quality. For example:

"Intense" → "I'm passionate and driven."
"Not good enough" → "I'm capable and improving every day."

"Incompetent" → "I am learning and continually growing in my skills."

"Boring" → "I'm steady and a calming presence."

The labels we wear—*too quiet, not leader material, always a perfectionist*—are the result of well-worn neural pathways, carved deeper each time we revisit them. The brain is efficient; it strengthens the connections we use most. So every time we reinforce a label, whether through self-talk, a memory, or others' expectations, it becomes more automatic, more ingrained, until it feels like the truth rather than just a story we've repeated. Let me explain why.

Mental Grooves: Carving New Paths to Overwrite Self-Doubt

When I was eighteen, I went sand dune surfing in Nelson Bay, a beachside town near Sydney. It was the ultimate end-of-exams celebration: sun-drenched summer, friends, and an endless wave of sand. A dune buggy hauled us up the hill, we got a quick tutorial, and they handed me a battered old sandboard. I wasted no time—I sat down, leaned forward, and started sliding. It was exhilarating!

But controlling the board wasn't as easy as I thought. About three-quarters of the way down, I hit a snag. My board dug into the sand and sent me flying face-first into a soft, powdery landing. I laughed it off, brushing grit from my mouth. Then came the grueling hike back up the dune under the scorching sun to do it all over again. After a few runs, I noticed something. The first trip down, my board carved out a fresh path. On the second and third runs, I started in the same place but veered slightly. By my eighth run, I had a clear, well-defined track—the "default" route my board naturally stuck to. Even when someone crossed the path and disrupted it, I could still find and follow the groove I'd created.

This is very much like how beliefs are formed in your brain. Every thought, every story you repeat to yourself, carves a pathway. Over time, those grooves deepen, becoming your automatic mental track. Your default. Your blueprint. Maybe even your "scar."

The more you repeat certain thoughts (whether it's self-imposed labels or confabulated worries), the more your brain locks them in as truth. This process happens through your brain's Automations Lead, which turns thoughts into default mental routines that are difficult to break. What starts as a passing thought gradually solidifies into a deep-rooted belief, shaping how you see yourself and the world. But just because your brain prefers to stick to these familiar grooves for the sake of efficiency doesn't mean you're stuck with them. What you do need, though, is a new way to think about them.

Remember: You weren't born with a self-doubt pathway etched into your brain. When you were a child, self-doubt wasn't part of the equation. You didn't question your worth or hesitate to try. You didn't worry about whether you were good enough to walk; you just kept falling until you figured it out. But somewhere along the way self-doubt crept in. Maybe it started in school when you were picked last for dodgeball and thought, *I guess I don't belong.* Or when your parents casually compared your grades to your sibling's, subtly implying that you weren't measuring up. Each moment was a slide down the self-doubt dune, deepening that groove in your brain.

Forget all the self-help clichés about "deleting old programs" or "erasing limiting beliefs." You can't simply delete or ignore these well-worn ways of thinking. They're habits that have been reinforced for decades. Neuroscience makes it clear that old neural pathways don't get erased. But they don't have to be. You can carve new ones right over the old tracks. The more you use the new path, the deeper it gets, and eventually it becomes your brain's new default.[7] Just as those doubt pathways were formed, they can be overwritten. Thanks to neuroplasticity (your brain's ability to adapt), you can create new, healthier pathways. It takes deliberate effort, repetition, and time, but it's absolutely doable.

Overwriting Your Mental Grooves

Think about those moments when something stressful hits at work or at home. What's your automatic reaction? Do you freeze? Start spiraling into self-doubt? Procrastinate like it's your full-time job? Those responses aren't random. They're the result of years of repeated thoughts and behaviors carving out well-worn neural pathways. Every time you think, *I can't handle this* or *I'm not good enough,* you're deepening those grooves, making it easier for your brain to go straight to Doubtsville the next time something tough comes up. But as we just learned, you're not stuck with them.

When Rashida reframed "intense" as "passionate," that allowed her to own who she truly was. As a result, she started showing up differently. She had more charisma, more confidence, more self-assuredness. But, in certain situations, her well-worn mental grooves still found a way to trigger her self-doubt. In meetings, especially virtual meetings, she'd freeze. When she felt all eyes on her, her brain would drag her back to childhood memories of being teased, and self-doubt would flood back in.

I challenged her to do something *different*. I wanted her to start creating new grooves in her mental sand dune to overwrite the old ones. I told her, "Next time that anxiety hits and you start to doubt yourself, do the opposite of what your body wants. Instead of retreating—looking away, slouching, leaning back—lean in. Literally. Sit on the edge of your seat, straighten your posture, take a deep breath, and smile. Physically signal to your brain that you're safe. Then, remind yourself, 'I'm here, I'm present, I'm safe, I'm showing up.'"

Rashida almost laughed when I suggested this. "You really think that's going to work?"

I didn't flinch. "I challenge you to try it. See if it makes a difference."

And she did. The next time her anxiety flared during a Zoom meeting, she short-circuited her conditioned response and used that as her cue to lean in—both physically and mentally. She adjusted her posture, smiled, and silently said to herself, *I'm here, I'm present, I'm safe, I'm showing up.*

It felt awkward at first, almost unnatural. But Rashida stuck with it. Over time, her new response started to take root. Each time she leaned in instead of pulling back, something shifted. She felt a little more in control. Her self-talk steadied her nerves. Plus, straightening her posture seemed to help her think more clearly and feel more confident.

There's a reason for this. It's called *embodied cognition*.[8] Our bodies and minds are connected. How you carry yourself can influence how you think and feel. Stand tall, and your brain starts to believe you've got this. Smile, and your stress starts to take a backseat. It's not magic. It's biology.

By the third month, Rashida didn't even have to think about it anymore. Leaning in and repeating to herself, *I'm here, I'm present, I'm safe, I'm showing up,* had become second nature. She'd created a new groove in her mental sand dune. Her doubts still popped up from time to time, but they didn't run the show. Over time, she found herself more present and engaged, and her confidence grew. A year later, she was promoted.

Rashida's commitment to rewiring her mental grooves paid off. She proved that with intention and consistency, she could rewrite the mental scripts that held her back.

PRACTICE: Create New Tracks—Take an Opposite Action

The next time self-doubt sneaks in and urges you to hold back or shrink, do the *opposite.* Pick something low risk, something manageable.

- If self-doubt tells you to shrink, slouch, or look away, notice it, and gently choose to open up instead. Expand. Take up space. Shoulders back, chin up, and hold someone's gaze even if for a moment.

- If your instinct in a meeting is to stay invisible, push yourself to speak up, even if it's just to say, "I agree with that" or "That's a great point."
- If you typically avoid asking for help because you don't want to look incompetent, try something simple: "Could you clarify that for me?"
- Set your own commitment: "When self-doubt makes me want to________, I'll choose to ________ instead."

What you're doing here is breaking the self-doubt autopilot and the mental habits that have been reinforced over time. This practice is rooted in the therapeutic approach of dialectical behavior therapy, where you do the opposite of your emotional urges when they don't serve you well.

If your emotional instinct is to retreat, step forward instead. If you feel like hiding, show up. The goal isn't to fake confidence or suppress what you're feeling, but to align your actions with how you *want* to feel, not how your doubts are telling you to feel.

• • • •

Beliefs are default thinking patterns that deepen each time you revisit the same story. Over time, they create mental grooves, guiding your thoughts even when they don't reflect reality. The labels you collect along the way act like signposts, reinforcing those grooves and pointing you back to the same narrative. But grooves can be overwritten, and labels can be peeled off.

And that has to happen for real growth to begin.

Every time you challenge the story you tell yourself about who you are, you disrupt the cycle. You create space—for growth, for possibility, for a different way of seeing yourself. Because your self-image isn't fixed. It's something you've constructed over time, built from the narratives you've absorbed and replayed until they felt like truth. And the way you define yourself doesn't just shape your present; it sets the limits of what you believe you can become. These beliefs influence the risks you take, the dreams you allow yourself to pursue, and the ones you abandon before even trying.

CHAPTER 3

What Is My Doubt Profile Telling Me?

The Individual Origins of Self-Doubt

So now we've reached the final part of the self-doubt story to give you a fuller understanding of the Doubt Profile—and why it matters.

The night I fired off that late-night question where I asked researchers, psychologists, professors, and coaches about the most effective way to sabotage someone's success, I didn't expect the answer to be so clear. Self-doubt was the common thread holding people back. And once I saw that, I couldn't look away.

Self-Doubt as a Crisis of Self

As I dived into more than fifty years of research on self-image and personal growth, I started to pick up on something I hadn't recognized before:

Self-doubt is not only a confidence problem. It's an *identity* problem. A crisis of self.

Even the term *self-doubt* breaks down to *self* + *doubt,* meaning you're not questioning just your skills or knowledge, but *your self.* Your value. Your place. Your right to take up space. You doubt your very sense of who you are. And that's why self-doubt sticks. Because we mistake it for *who we are* rather than something we've learned or internalized.

With this guiding idea, I went back and analyzed hundreds of client cases, and I saw how this pattern of "self-doubt as identity" played out across industries, roles, and personality types. People tended to interact with self-doubt in one of two ways:

- The first group, which struggled most with doubt, not only *felt* self-doubt, but *engaged* with it constantly. They *became* it. Their instinct was to "fix" the self-doubt, to treat it like a problem to solve, a flaw to correct. They tried to analyze their way out of it, self-improve their way through it, and criticize themselves into getting better. But the more they focused on "fixing" it, the louder it became.
- The second group felt doubt too, but they didn't let it define them. Instead of ineffective attempts to eliminate it, they *balanced* it. They leaned into other parts of themselves—inner strengths, values, and deeper truths—that steadied them. They anchored themselves in a deeper sense of *who* they are, beyond their insecurities.

It wasn't that the second group had *less self-doubt.* But they knew the doubt wasn't them. It was something they could peel off, burr by burr.

I wanted to test this. If self-doubt wasn't random, could we trace its patterns? Could we isolate the parts of ourselves that are vulnerable, and more importantly, strengthen them?

Over several years, I built on research from the fields of organizational behavior and personality psychology and studied more than 4,000 people—through research, interviews, and real-life case studies.[1] The data was clear: Self-doubt doesn't strike randomly. It finds and attaches to specific aspects of how we see ourselves—our sense of worth, our feeling of being capable and ready, our sense of ownership over our actions and choices, and our emotional groundedness.

That means that if you want to rise above your doubts, you need to build up the parts of yourself where self-doubt tries to undermine you.

That's what led me to develop the Doubt Profile[2]—a tool to pinpoint where doubt takes hold, and actionable ways to rebuild self-trust.

The four core Attributes—Acceptance, Agency, Autonomy, and Adaptability—shape how you experience self-doubt and how much power it has. Each one influences how you interpret setbacks, respond under pressure, and decide whether to move forward or stay stuck. Each Attribute reflects a fundamental question you ask yourself—often unconsciously:

- **Acceptance**: Do I believe I'm worthy as I am?
- **Agency**: Do I trust my skills and abilities?
- **Autonomy**: Do I feel I can shape my path?
- **Adaptability**: Can I stay emotionally grounded when doubt arises?

These Attributes don't operate in isolation either. They're deeply connected. When one is low, it can drag others down with it, giving self-doubt the perfect opening. But each Attribute can be strengthened—not by forcing it, ignoring it, or pretending doubt doesn't exist, but by shifting the small patterns you repeat daily. We can see our doubts as habits we default to and reinforce every day. They are the choices you make under stress and the way you show up when it counts. Change those patterns, and you don't just quiet the doubt, you change the way you fundamentally see yourself. That's how Big Trust is built.

The Doubt Profile in Practice

To see how revealing and helpful the Doubt Profile can be, let me share with you Johan's story. He was outgoing, talented, and full of energy. The kind of person you'd expect to thrive in any room. He had undeniable talent and drive as a graphic designer at a major tech platform.

But beneath the surface was a storm of second-guessing. Johan was never satisfied with himself. He worried constantly (though he was careful to never show it). He obsessed over every interaction, replaying conversations to analyze whether he'd said the right thing, contributed enough, or appeared competent. Despite his success, he carried an invisible scoreboard, tallying his perceived failures.

We'd been working together on some big career moves he was in the middle of—interview prep, professional strategy, the works. Then the moment came. He was offered a leadership role (exactly what we'd spent months prepping for). And he did the unthinkable. He turned it down. I was baffled.

"What's going on?" I asked.

Johan shrugged. There was no clear explanation, no specific fear he could name. But his decision spoke volumes.

As we dug deeper, it became obvious that his choice was a shield. It was a way to dodge the discomfort of stretching beyond his comfort zone. He wasn't afraid of the work or the responsibility—he was afraid of what it would confirm about him if he stepped up and failed. He kept thinking: *I'm not a leader. I crumble under pressure. I'll screw it up.* He saw his future self and thought, *That can't possibly be me.*

That's when I knew what we needed to do. If I could help Johan see how the Four Attributes were quietly steering his choices, I knew he could live up to his potential.

And here's what I realized: Johan wasn't struggling with clarity or skill or ambition. He was struggling with himself. Or more specifically, with how he *saw himself.*

With Johan, the patterns were easy to spot. Here's how his Doubt Profile looked:

- **Acceptance** landed squarely in the Red Alert zone. His self-worth was completely tied to achievements. If he hit a bump or made a mistake, it felt personal, like a verdict on his value. That kind of pressure kept him playing small. When Acceptance is weak, success always feels conditional, like it could disappear at any moment.
- **Agency** sat in the So-So zone. That meant it was inconsistent. Johan had self-assurance in familiar situations, but with new challenges or higher stakes, his self-doubt about his capability crept in fast: *What if I fall flat on my face? Do I really have what it takes?*
- **Autonomy** flagged another Red Alert. Johan didn't feel in control of his career, or even his identity. He felt boxed in, convinced that his path was controlled by his boss, the industry, or the economy. That left him feeling stuck, believing he had no say in his future.

- But **Adaptability**? That was a Hidden Strength. Johan had a rare ability to stay calm under pressure (which he learned from being the eldest of seven kids and playing "parent" from his early teens). It helped him function through stress, but it also masked deeper struggles with other Attributes (like Acceptance and Autonomy) that he never fully processed.

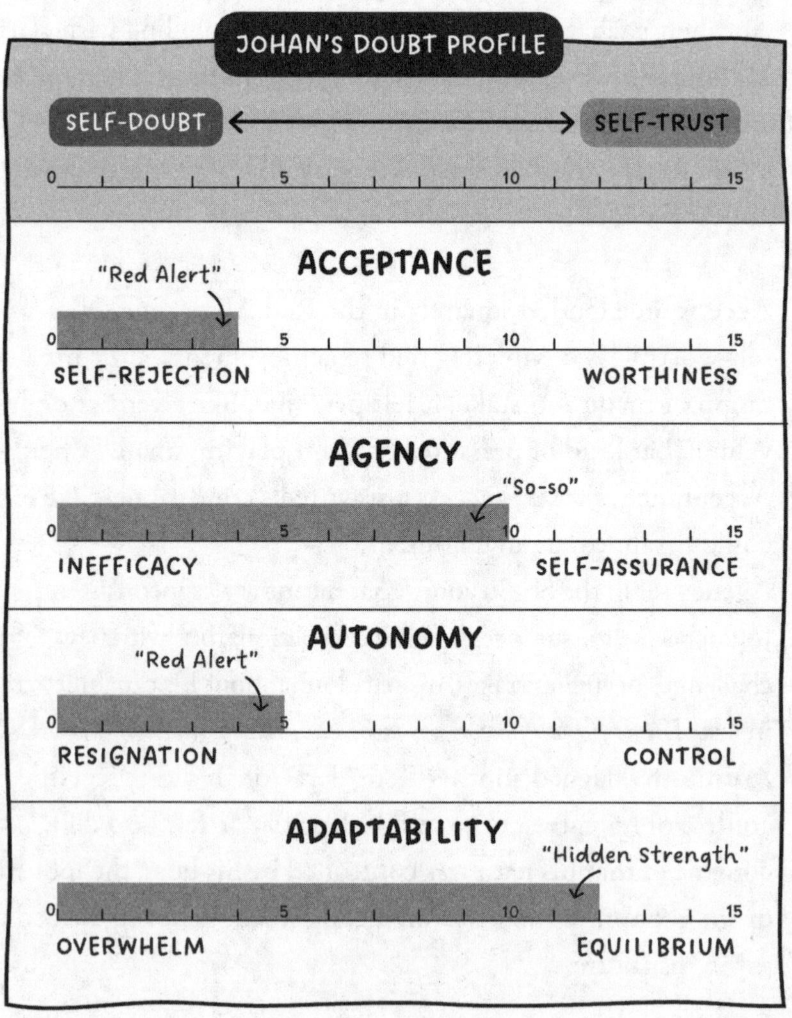

Once we mapped out these patterns, Johan's work became clear. There was no "fixing" involved, just patterns to shift and Attributes to strengthen so his brain and internal stories didn't fall into their default modes. We focused on simple habits that were tied to Acceptance and Autonomy as a priority. For Acceptance, the focus was on expanding his sense of identity beyond his job. He started a weekly reflection practice where he asked himself, "Where did I show up this week, outside of my core work responsibilities?" Sometimes the answer was "as a present dad," "as a thoughtful friend," or "as someone who brought calm to a hard situation." These reminded him of his worth beyond his job performance. For Autonomy, whenever things started to feel out of his control, he practiced asking himself, "What's one thing I can choose here, even if it's just how I show up?"

And while Johan wasn't exactly jumping for joy at the start (he was more skeptical than anything else), I could see the spark flicker to life as he began to experience things differently. Just over twelve months later he stepped into the leadership role he so deeply wanted—this time with the confidence built from the small habits he'd been practicing all year.

Widening the Possibilities

In the years since developing the Doubt Profile, Fayçal and I have tested this framework with Fortune 500 execs, entrepreneurs, creatives, and thousands of other professionals.[3] For many, it was like looking in a mirror for the first time and *really* seeing themselves. They'd say things like, "This explains so much," or "I finally get why I keep holding myself back."

The real power came when they started using the framework. They started developing the core habits each Attribute represents—

like the habit of accepting yourself or the habit of practicing emotional adaptability—and applied the tools and practices that bring those habits to life.

They worked to strengthen the areas where doubt had the strongest grip, and reinforced the parts of themselves that were already solid. Their overthinking and second-guessing started to fade, their anxiety became less overwhelming, their hesitation gave way to bold, intentional action.

But the biggest shift was how they began to see themselves differently. Not as imposters. Not as people stuck in their limitations. But as capable, grounded, courageous humans with untapped potential. They were becoming better leaders, better parents, better partners, and better versions of themselves. They were braver in *every* area of their lives.

In fact, by the time we wrapped up our work together, Johan told me, "I love this process. I leave with real, concrete steps I can take, and I'm already seeing the changes. My team, my clients, even my friends have noticed a difference." Johan had stepped into a bolder version of himself. He'd taken ownership, taken action, and he saw the impact in his work and life. He leaned into his strengths and did the work to reinforce the areas that needed more support. And that's how he began to build Big Trust—one decision, one action, one habit at a time.

And you can do that too.

PRACTICE: Mapping Your Self-Doubt Starting Point

This practice will help you reflect on what's holding you back and commit to the process of reshaping your self-image to move forward.

Step 1: Review Your Doubt Profile

Look at your Attributes—both your strengths and the areas flagged as challenges. List your score for each of the Four Attributes:

Acceptance: ___
Agency: ___
Autonomy: ___
Adaptability: ___

Step 2: Spot the Patterns

Now that you've seen where doubt shows up, it's time to explore how it plays out in your life. Sometimes your self-doubt will be obvious, but other times it quietly operates in the background. Both types matter:

Daily doubts chip away at your confidence over time. Think of moments like hesitating before sharing an idea, second-guessing an email before hitting send, over-apologizing when it's unnecessary, or ruminating over a misstep (or potential misstep) and not being able to move forward.

Big doubts are the defining ones—the moments that feel like they'll shape your future. Turning down an opportunity. Not applying for the job you really want. Staying silent or not asking for help when raising the issue could have changed the outcome.

Reflect on both as you answer these questions:

- What's holding you back—old beliefs, labels, or habits?
- How are daily and recurring doubts affecting your actions?

- When has self-doubt shaped a big decision, and how has that shaped your life?
- In your daily actions, when you hesitate, hold back, or second-guess yourself, is there a primary Attribute that might be at play?
- Are there any Attributes that are actually strengths you've overlooked because doubt has been louder?

Step 3: Commit to the Process

Commitment is showing up for yourself, even when doubt whispers that you can't, you shouldn't, or you don't deserve to. Commit to the process of strengthening your Attributes, one small, deliberate action at a time. Even strong Attributes can be fortified further.

Say this commitment out loud or write it down:

I commit to showing up for myself, to challenging the doubts that have held me back, and to strengthening the parts of me that are ready to grow. I'm not "fixing" who I am, but uncovering who I've always been beneath the doubt.

Grow Your Self-Trust Habits

You're standing at the edge of who you are now and who you *could* be. Our goal is to bridge the gap between the *you* that talks about what you want and the *you* that actually goes out and gets it. You'll trust yourself more, doubt yourself less, and show up in a way that actually feels like *you*. You'll take action with more courage, creativity, and enthusiasm.

Because the more doubt you carry, the more your capacity for growth shrinks. Not because your potential has changed, but because your access to it has. That's what self-doubt does. It convinces you that the walls around you are real, when they're just reflections of outdated beliefs and invisible limits you've internalized.

• • • •

I didn't fully understand this until I saw it play out *literally* in my own living room.

When Fayçal and I first relocated to Southeast Asia (our halfway base between our families in Australia and our clients in Asia, Europe, and the US), we decided our new home could use some tropical flair. So we bought a six-and-a-half-foot Manila palm tree and plopped it in the living room.

Fun fact: If planted outside, that tree could grow to twenty-five feet. But indoors? Stuck in its little pot? Six feet was its limit. Not because of its nature, but because of the pot.

Now, imagine if that tree had conscious awareness. Imagine it started to wonder why it never measured up to the trees outside. It might think, *This is all I am—a six-foot tree.* But its size isn't a reflection of its potential. Move it to a bigger pot, and it grows taller. Plant it in open soil, and it shoots up to its full height, effortlessly becoming what it was always capable of—just waiting for the right space to grow.

Dutch author Alexander den Heijer nailed it with this quote: "When a flower doesn't bloom, you fix the environment in which it grows, not the flower."

Except the pot isn't physical. It's made up of the mental boundaries you've built—labels you've internalized, stories you've repeated, and beliefs you've never thought to question. Self-doubt makes you mistake the pot for your potential. That the limitations you've learned are fixed. That the voice of doubt is who you are.

But it's not.

Your doubts were never proof of your limits. They were just the edges of a pot you were never meant to stay in.

The real shift begins when you recognize that the edges of that pot are shaped by the habits you practice every day. When you change your habits, you expand the space you live in. You stretch. You grow. And doubt begins to loosen its grip.

In the next four parts of the book, we'll unpack each of the Four Attributes (yes, they matter, and no, I don't recommend you skip them), and by the end of it, you'll start seeing yourself differently—less "stuck" and more "heck yes, I can do this."

You'll have the proven, practical tools to pause, think, and act in ways that actually move you forward instead of keeping you spinning in place.[4]

Because doubt doesn't define your limits. *You* do.

And it's time to stop standing at the edge of who you could be—and start living it.

The First Attribute

ACCEPTANCE

CHAPTER 4

Am I Enough?

The Core Question of Acceptance

Back when my career was just getting started, I was a chronic over-apologizer.

"Sorry for rambling . . ."

"Sorry, did that even make sense?"

"Sorry for saying sorry so much . . ."

At one point, I probably would've apologized for existing if I thought it would make people like me more. At the core, I felt like that piece of gum on the sole of someone's shoe. An annoyance. An inconvenience. The voice in my head had me convinced that I was the equivalent of background noise—less important, less insightful, just *less*.

And it showed. People's reactions to me weren't subtle. Some would quickly reassure me with a half smile, but with a hint of impatience, as if my apologizing was *interrupting* the conversation. Others would just nod vaguely, eyes already glazing over, as though they were mentally checking out. It was brutal, and it confirmed what that little voice in my head had been screaming all along: *You don't matter.*

I started carrying that energy into meetings, prefacing every idea with "Sorry if this is a dumb idea." Guess what happens when you do that? People start assuming your idea *is* dumb—because you told them it probably was. I was undermining myself before I even had a chance to make my case. By constantly apologizing, I was almost giving everyone around me permission to see me as small, insignificant, and not worth taking seriously.

Despite all this, I somehow managed to get very good at *looking* like I had it all together. On the outside, I was achieving, ticking boxes, moving forward. Inside though, I felt like I was pretending, and that feeling weighed on me.

Four years of grinding away as a paralegal at a top-tier firm while finishing my law degree left me burned out, drained, and physically wrecked.

So, I left.

Following my dad's advice (and my total lack of other plans), I switched careers entirely, jumping into retail banking with one of Australia's Big Four banks. My dad had worked there in the commercial team for years and always spoke highly of it, so I applied and was fortunate enough to get a spot in a special training program in another area of the business. (I'll never forget calling him and saying, "Guess what, Dad? See you in the office on Monday!").

I thought leaving law would be a fresh start. That my doubt could be left behind with my employment pass. But no, the burrs of doubt are way more clingy than that. It followed me, repackaged for the new environment. Every unfamiliar acronym, every high-stakes meeting, every new expectation carried the same whisper: *You're not enough.*

So, I overcompensated. I chased validation through late nights, extra projects, saying "yes" to everything, relentless perfectionism—hoping it would finally quiet that voice in my head.

It didn't.

The harder I worked, the emptier it all felt. Turns out, you can't outrun the feeling of unworthiness.

Why Worthiness Matters

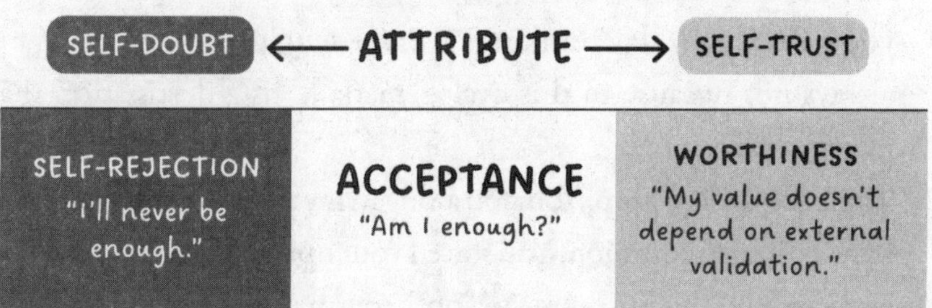

The search for worthiness is the defining challenge of Acceptance, the first Attribute in your Doubt Profile. When Acceptance is a vulnerability, self-doubt turns inward, and you begin to reject yourself. You criticize yourself before anyone else gets the chance. You chase perfection to outrun the fear of being exposed. You try to fail-proof your life just to avoid the shame of "not enough." Every flaw feels like proof that something's wrong with you. Every success is downplayed or dismissed.

Not everyone who struggles with self-doubt has challenges with Acceptance. But our research shows that 54 percent of people fall into the Red Alert or Hindrance zones—meaning it's either actively disrupting their lives or quietly holding them back without them even realizing it. Another 20 percent land in the So-So range.[1] That's nearly three out of four people walking through life with a fragile sense of worth. What does this mean in real life? When the habit of Acceptance is weak, it tends to surface in four painfully familiar patterns:

1. **The Pressure to Prove.** The brain's response to feeling "not enough" is often to overcompensate.[2] You tell yourself that the next achievement, promotion, or milestone will be *the one*—the moment you finally feel like you belong. But the finish line keeps moving. You take on more, chase perfection, and tie your worth to your output. Feedback feels like a personal attack. Compliments? You dismiss them. No matter how much you achieve, you still end the day thinking: *Was it enough?* And the answer is always no. Because in this cycle, "made it" is a destination that never arrives.
2. **The Likeability Trap.** You nod along in meetings even though you have a different opinion. You soften your opinions, over-apologize, and say "yes" when you mean "no." You make yourself easy—easy to work with, easy to be around, easy to like. You settle—for jobs, for relationships—because asking for more feels risky. What if you push too hard? What if you *become* too much? You pour yourself into others, hoping that this time, it'll be enough. That *you* will be enough. But it never is.
3. **The Shrinking Syndrome.** Fear of success can be just as paralyzing as fear of failure. Not because success itself is scary, but because the more visible you become, the more pressure you *imagine* is on you. More eyes, more expectations, more chances to disappoint. So, you procrastinate, you hesitate, and you convince yourself you're fine where you are. But you're not. You're just scared. This is what psychologist Gay Hendricks calls the Upper Limit Problem[3]—where safety feels better than growth. But staying small isn't safe. It's self-sabotage.
4. **The "Schadenfreude" Cycle.** When you're deep in self-doubt and aching for a sense of worth, someone else's misstep can feel like a strange kind of comfort.[4] It's not that you're a cruel

person—it's a momentary escape from your own inadequacy. But that ego boost is a sugar rush, fleeting and empty (and you might feel a twinge of guilt after). Watching others struggle won't make you feel whole. It only keeps you stuck in comparison, reinforcing the very insecurity you're trying to escape.

These patterns—overworking, people-pleasing, shrinking, or finding comfort in others' failures—all stem from the same belief: *I'm not enough.* Every behavior is an attempt to avoid that discomfort. But until you face it head-on and call it out for the lie it is, you'll stay stuck.

From Scarcity to Enoughness

The challenge of "enoughness" doesn't start in adulthood. It's been trailing us since before we had words for it, shaped by every hug, every smile, every moment of reassurance (or lack of it) in our earliest relationships.[5]

If love felt conditional—tied to achievements, behavior, or meeting high standards—you may have developed what developmental psychologists call anxious attachment.[6] From that foundation, a belief quietly forms: *I have to earn love. I have to prove I'm enough.*

This attachment blueprint follows us into our adult lives, fueling what I think of as a scarcity outlook, where worthiness feels limited. It becomes something you must earn rather than something you already possess. The search for worthiness turns life into a competition where you fear failure and chase approval and perfection. You become trapped in a cycle of needing more: *more* validation and *more* proof that you're enough. You fixate on what you lack: not smart enough, not attractive enough, not hardworking enough, not funny enough, not fit enough, not educated enough, not successful enough. A single critique feels like confirmation that you don't measure up. Someone else's success feels like proof of your own inadequacy.

On the other hand, if you grew up in an environment where love and validation were freely given, you likely developed what's called a secure attachment, a foundation for feeling safe in your own skin, a steady sense of self-worth in adulthood. Research backs it up: Kids who feel securely attached tend to get better grades[7] and grow into adults with higher self-esteem,[8] better relationships,[9] and frankly, fewer emotional breakdowns over things like accidentally sending a text to the wrong group chat. This creates an "enoughness" outlook—a deep trust that you are inherently worthy. When you live in *enoughness*, success isn't a limited resource, and your value isn't up for negotiation.[10] You stop seeking permission to take up space. This deep, unshakable trust in your own worth is the foundation of acceptance.

Think about it for a moment. Do you find yourself stuck in thoughts of scarcity, constantly proving, comparing, and chasing? Or are you rooted in enoughness most of the time—trusting in your worthiness, regardless of your achievements?

Most of us don't land squarely in one camp. We ride a seesaw—some people and situations can trigger old scarcity patterns and flood us with self-doubt, while others bring out our inner trust in who we are and make us feel that we're enough. And the reasons we ride that seesaw aren't always obvious. They're shaped by not just what happened, but also how we interpreted it, what we made it mean about us.

And sometimes, what you made it mean doesn't line up with what is actually true.

I grew up in a loving home, with plenty of support. My parents split when I was fifteen, but even then, I had a steady flow of love and affirmation from both sides. By all accounts, I should have had a rock-solid sense of self-worth. And yet, I didn't.

Beneath the surface, a quiet voice whispered: *Don't mess this up. You're only as good as what you achieve.*

I wasn't just trying to make my parents proud; I was terrified of disappointing them. I tied my worth to being the "perfect" daughter, the overachiever, the one who never let anyone down. The one who never made life harder for anyone else.

And then came the guilt: *What's wrong with me?* I had no reason to feel this way. I had love, opportunity, stability—everything that should have made me feel secure. But that only made the weight heavier. Instead of questioning the pressure itself, I questioned myself. *Why can't I just be grateful? Why do I still feel like I'm not enough?*

Here's what I wish I had known sooner: Nobody is 100 percent secure in their self-worth all the time.[11] We all have moments of questioning, of tying our value to something external. Welcome to being human.

The good news is that whatever early experiences you had and whatever your current state, you're not locked in. Hundreds of studies prove that even if you grew up in an environment that reinforced an attachment style based on a scarcity outlook, they've simply become habits of thought—and that means they can be replaced with new ones.[12]

Understanding the foundations of your own Acceptance helps you see where self-doubt takes hold so you can replace those reflexes with something stronger: the belief that you are *already* enough, even as a work in progress.

The Risk of a Single Identity

Most people struggling with self-worth find themselves clinging to a single identity—"I'm a lawyer," "I'm an entrepreneur," "I'm the responsible one." It's a way to feel secure, to anchor a sense of *enoughness* in something recognizable. But this is problematic. When your entire sense of self is tied to one role, any setback—a bad review, a layoff, a mistake—can feel like a personal failure. Your very identity and sense of self is called into question.

Cognitive scientist Dr. Maya Shankar calls this process Role Fusion. She learned it the hard way. As a child she was a violin prodigy, accepted into Juilliard and handpicked to study under one of the world's top violinists, Itzhak Perlman. But at age fifteen, during a summer music camp, she woke up early to practice, didn't warm up properly, and overstretched a finger on a single note. *Snap*. A torn tendon crushed her dreams of ever playing professionally.[13] In her TEDx Talk, she admitted, "I found myself grieving not just the loss of the instrument, but also the loss of myself. For so long, the violin had defined me, that without it, I wasn't sure who I was or who I could be. I felt stuck."[14]

What we've noticed among the *top* performers (who are also the happiest) is that they have something to focus on outside of work—a passion, a creative outlet, a hobby. It gives them an identity that isn't tied to productivity, success, or anyone else's expectations.

I learned this firsthand in my mid-twenties.

When I transitioned from law to banking, I felt like I was constantly five steps behind, overanalyzing everything, terrified of making the wrong move. It felt like everyone else spoke a language I hadn't learned yet, and I was scrambling to keep up. Then one Monday night, my sister-in-law invited me to join her and my older brother Ryan for a Latin dance class. I'd danced as a kid, but like a lot of things, I'd left it behind after high school. I had no idea how much that single class would change me. It was the first space where I wasn't optimizing, strategizing, or chasing achievement. I was just showing up to enjoy the experience.

In one night, I got hooked. I threw myself into lessons, started teaching and, soon, competing. My dance partner and I won three national titles. I even represented Australia as a soloist at the World Salsa Open in Puerto Rico (that was fun). For once, I wasn't overthinking—I was fully in it.

But here's what was most interesting. Dance didn't just give me confidence onstage; it changed how I showed up *everywhere*.

I stopped obsessing over being "flawless" in every meeting and started focusing on adapting in real time, just like I did on the dance floor. Instead of second-guessing every decision, I trusted my instincts more—because dance had trained me to think on my feet. I stopped seeing mistakes as failures and started seeing them as feedback, just like in training sessions. If I could get back up after missing a step in a routine, I could handle a misstep in a business deal.

At work, I started channeling the version of me who could step onto a competition stage under blinding lights—focused, instinctive, trusting myself. And it worked. That boldness carried over. I stopped bracing for failure and started moving with confidence. And when work drained me, I had dance training to look forward to—reminding me I was more than my job.

When your identity extends beyond your job, or your title, your self-worth stops living and dying by your performance. You don't break when things go wrong—you bend. You remind yourself that you're not just a "consultant," a "leader," or a "parent"—you're also a musician, a climber, a painter, or someone who loves to transform thrift-store furniture into works of art they sell on Etsy.

This isn't just feel-good advice. It's grounded in hard evidence. In a study published in 2023 with more than 93,000 participants across 16 countries, the researchers found that having hobbies was linked to better health, greater happiness, and higher life satisfaction.[15] Other studies have found that people who pursued hobbies unrelated to their work even felt more confident at work,[16] and having hobbies can boost your sense of worthiness and self-esteem.[17] Even something as simple as following a recipe to cook a meal[18] or taking a dance class[19] (I can vouch for this one firsthand) can make a measurable difference in how you see yourself and your identity.

Hobbies reinforce self-worth because they remind you that your value isn't just about what you produce (or who you please). And when life throws setbacks your way—a missed promotion, a bad review, a burned turkey in the oven because you forgot to set the timer—you don't crumble, because your sense of self isn't precariously riding on just one thing. Your identity is broad, balanced, and deeply yours.

Think about it. In your own life, are you currently a runner or on a basketball team? If not, was there a time when you weren't just the "you" who works, parents, or overthinks? Were you a baker once? A volleyball player? Did you love photography, gardening, or playing the drums? How did you feel? How did you interact with others (or with yourself)?

Who was that version of you?

Imagine for a moment what it might look like to bring a little of that back.

And if nothing comes to mind, maybe it's time to pick up something new—an outlet, a passion, a trial endeavor that's just for you. The goal here is to expand your identity and rediscover the parts of you that are not tied to work, responsibilities, or what others expect. When you reconnect with that "other you," you start to see yourself as more whole—and far less lacking.

The Promise of Acceptance

When you accept yourself, it shows up in how you move through the world and how you relate to yourself. You feel a grounded sense of being "enough" and know you are valued for who you are.

Self-acceptance gives you that steady presence—the ability to sit in a tense meeting and speak with calm confidence, without questioning whether your input matters. It's what allows you to listen to your significant other without spiraling into self-blame or assuming you've done something wrong.

With acceptance, you can have a bad day, face a setback, or navigate conflict without making it a verdict on your worth. Instead of getting stuck in a cycle of doubt, you find the resilience to regroup, reengage, and focus on what comes next.

You stop outsourcing your worth. You no longer wait for others to approve of you before giving yourself permission to feel capable or deserving. And while you embrace who you are right now, you're also kind enough to yourself to want to grow. That kindness fuels your motivation,[20] not out of shame, but out of care.

Most of all, you create a sense of inner peace—the kind that stays with you no matter what the world throws your way.

Your ability to accept yourself—flaws, quirks, and all—shapes how you show up in life. When self-doubt takes the wheel, success, love, and happiness feel just out of reach, like they're reserved for someone else. But the moment you stop questioning your worth at every turn, you start dismantling the walls that doubt built. You begin to see what's been true all along: You are enough, right now, as you are.

PRACTICE: Assess How Acceptance Is Showing Up for You

When you understand the role Acceptance plays in your Doubt Profile, you get a clearer picture of where those beliefs have come from. You begin to see where you're most vulnerable and the internal stories and habits that make you question your worthiness. But more importantly, you start to see how self-acceptance can be built—one shift at a time—until Big Trust no longer feels out of reach, but like something you can return to, again and again.

What Is Your Acceptance Attribute Score?

Look back at your Acceptance score from the Doubt Profile diagnostic. What was it?

Acceptance Score: __________ Zone: ____________________

(For Zone, write in "Red Alert," "Hindrance," "So-So," "Hidden Strength," or "Superpower.")

Use these prompts to dig deeper. You might like to journal your responses:

What are your vulnerabilities around the search for Acceptance?

- **Think back to childhood—what messages did you absorb about self-worth?** Were you taught that you were enough just as you were, or did you feel like you had to prove yourself?
- **How does self-rejection show up in your life?** Do you downplay your achievements? Brush off compliments? Keep people at arm's length because deep down, you don't believe they'd accept the real you?
- **When was the last time you truly felt accepted—for exactly who you are?** What was different in that moment? Who were you with? And what would it look like to extend even a fraction of that same Acceptance to yourself?
- **How might your life change if you accepted yourself as you are, right now?** What would open up for you?

When self-acceptance feels impossible, what strength can you lean on?

- If you struggle with feeling "enough," you can rely on and develop your **Agency,** which is your belief in your skills and that you can take action, even if you don't feel ready. You may not feel like you're enough, but you can act like someone who keeps going anyway. What's one action you can take today to remind yourself of your capability? (You'll learn more about Agency when you get to the second Attribute.)

- If you're caught up in external validation, you can turn to and cultivate **Autonomy,** which is your internal sense of control. You may not control every situation, but you always have control over your next move. That's what keeps you from feeling powerless. What's one choice you can make today that reminds you you're not stuck? (You'll learn more about Autonomy when you get to the third Attribute.)
- If your emotions feel overwhelming, anchor yourself in your **Adaptability,** which is your capacity to regulate and recalibrate. Take a breath, create space between reaction and response, and remind yourself that no feeling lasts forever. (You'll learn more about Adaptability when you get to the fourth Attribute.)

• • • •

In chapters 5 through 8, you'll uncover the specific thought and behavior patterns that weaken your Acceptance Attribute and learn how to rebuild it. We'll expose three of the biggest barriers to self-acceptance: the critical voices that hold you back, the masks you wear to fit in, and the fear of imperfection that keeps you hesitating. Then, in chapter 8, we'll step back to explore a powerful way to shift your perspective and release the feeling of not being enough. It's something you can return to anytime, anywhere, to reconnect with a deeper sense of self-worth.

CHAPTER 5

Should I Believe the Voices in My Head?

Talking Back to Your Inner Deceivers

For one of my first-ever major presentations, I was invited to deliver a keynote on peak performance. The client was a private, invitation-only leadership organization for young chief executives. After a briefing call with the organizer, I couldn't have been more excited. What could be better than speaking to sixty emerging leaders at an exclusive dinner event?

I arrived at the venue, ran through my slides, got mic'd up, and felt ready.

As I walked onto the small stage, I could see at least twenty pairs of folded arms—that dreaded sign of resistance.

That's when I noticed it. The youngest person there must have been fifty. I'd assumed *young* meant, well, *young.* I never thought to ask their ages during the pre-event briefing call. It turns out, I wasn't addressing fresh-faced start-up founders like I'd thought. I was standing in front of the *elite* group of CEOs, who had started their CEO journey more than a decade earlier when they were under forty-five. They were far older than I'd anticipated, more experienced, and, as I quickly found out, radiating skepticism.

I took a breath and began, but I was instantly uncomfortable. As I pressed forward, I felt the growing weight of a room that wasn't with me. My brain filled the gaps: *They think I have no idea what I'm talking about. They're wondering why I'm even here.*

And that's when I heard it—the voices in my head.

One voice scolded: *Why did you say it like that? That was stupid.* Another whispered: *It's not too late to cut this short and get out of here.* Then came the one that always demanded more: *You're not trying hard enough. Throw in more facts. Be more authoritative. Prove you're credible!* And finally, the voice that begged for approval: *Change your style of delivery. You need to win them over.*

In real time, I scrambled to adjust. I leaned in harder, trying to sound more impressive. I got more animated, trying to turn their crossed arms into a semblance of engagement, or even a nod or two. I started speaking faster, layering on statistics and jargon, trying to impress them with my knowledge instead of connecting with them through genuine insight. My delivery became stiff and unnatural as I tried to mold myself to whatever I thought they wanted. It felt awful.

The talk ended. The host was thrilled (he had booked me, after all), but I knew. A large part of the audience wasn't convinced. As I walked back to my hotel that night, my mind replayed every misstep. *What a disaster. You messed that one up. The client is going to trash-talk you to your speaking agent, and there goes your speaking career.*

I had faced tough audiences before. But this was different. Because the real battle hadn't been against skepticism in the room; it had been against the voices in my head.

I had heard those voices many times before, but that particular evening, they were so loud in my mind that I couldn't shake them—and I couldn't be myself.

All of us have an internal dialogue running in the background, shaping how we see ourselves and the world. It helps us problem-solve, plan, and navigate life. But when the voices turn against us—when our inner voices fuel doubt instead of clarity—they become something else entirely.

Performance psychologist Dr. Jim Loehr has worked with hundreds of world-class performers and has dedicated years to studying the self-talk of his elite-level clients, trying to get inside their heads. He even had athletes wear microphones and vocalize everything they said to themselves in the heat of competition. He shared, "I began to realize what really mattered, in a really significant way, was the tone and the content . . . of the voice that no one hears."[1]

Those voices that no one hears shape your very identity.[2] Because whatever your inner voices tell you, you start to believe—and then you *become*.

Meet the Four Inner Deceivers

Between 2018 and 2020, before starting my PhD, I conducted hundreds of in-depth interviews with professionals to explore how people navigate their internal battles and break through self-imposed barriers.

What emerged was a strikingly common theme: More than three-quarters wrestled with the same relentless thought: *I'm not enough.* As I analyzed the transcripts, I saw these voices weren't just self-doubt; they reflected deeper struggles with worthiness and self-acceptance.

Psychologists call it the "inner critic," but my research revealed it's not just *one* voice—it's a whole squad of Inner Deceivers (aka, the "Jerk Squad"). I identified four distinct archetypes, each with its own flavor that distorts our self-perception and sabotages our self-trust.[3]

Here's the sneaky part. These voices don't show up looking like saboteurs. They *sound* like they're trying to help—offering what seems like protection from failure, rejection, or insignificance. But they're terrible at their job, and the security they offer comes at a steep cost: playing small, second-guessing yourself, and endlessly questioning your worth.

When I introduced these voices in my 2021 TEDx Talk, the response was overwhelming. It became one of the year's most watched TEDx Talks[4] and was selected by the TED team as an editor's pick—and I believe for good reason. These voices are universal. And left unchecked, they become impossible to quiet. That's why separating them, naming them, and calling them out is so powerful. It gives you a new kind of clarity.

For years now, clients and workshop participants have described their *aha* moments when learning about their Inner Deceivers. Once they see them for what they are, the lies start to unravel. And when you challenge those voices, they lose their power over you.

So, let's meet the squad:

THE FOUR INNER DECEIVERS

The Classic Judge

Do you have a mind that just won't quit judging you—constantly replaying what you've done, what you didn't do, or what you *should* have done differently? That's the Classic Judge at work. It just can't let things go. It latches onto your past mistakes or perceived failures, trapping you in a cycle of self-criticism. This inner voice doesn't just point out where you might have gone wrong; it magnifies those moments, making them feel like undeniable proof that you're not enough. It's that relentless whisper that says, *Why did I do that?* or

I should have handled that differently, pushing you into a spiral of overthinking and harsh self-judgments.[5]

The Classic Judge thinks it's protecting you from future mistakes by keeping you hyper-aware of past ones. It thinks that if it punishes you enough, you'll never mess up again. But it doesn't make you wiser. It just keeps you stuck in shame.

One of the research participants I interviewed was Felix, a chiropractor with a lifelong dream of writing a crime novel, or even a crime series. He had the main character mapped out in his mind for years, and he'd sometimes imagine scenes while he was working silently with his patients. But every time he sat down to write, his Classic Judge kicked into overdrive, nitpicking every sentence. He'd delete entire paragraphs and start from scratch, not because his writing was bad, but because he was stuck in this belief that his work wasn't worth reading.

Yet despite all that, Felix didn't throw in the towel. In a follow-up research interview three years later, Felix shared that he'd finally finished the novel, found a literary agent willing to represent a first-time author, and had landed a modest book deal. The book even became an Amazon bestseller, yet his Classic Judge still wouldn't let him enjoy his success. It kept pointing out flaws, making it hard for him to feel truly satisfied. It's a reminder that no matter how much we accomplish, the Classic Judge can still have us questioning ourselves, keeping that sense of "not enough" alive.

The Misguided Protector

The Misguided Protector is the paranoid voice that torments you with a worrisome future that hasn't happened (and probably never will). It loves to plant seeds of doubt in your mind about your ability to handle failure or rejection.[6] It's risk-obsessed, always urging caution: *You're not ready, you don't know enough, you're not qualified enough. You're too young. Too old. Too unprepared. Better not risk it.* It's basically a professional catastrophizer.

Its intentions aren't malicious—it's trying to shield you from pain. But in its effort to keep you safe, it also keeps you small. What starts as protection quickly becomes paralysis.

This was the main issue for Sandra. Full of innovative ideas and with a big vision for a start-up, her head was constantly filled with doubts about the future by her Misguided Protector, whispering, *Are you really sure you can manage a business? It's going to be a colossal failure.* This Inner Deceiver paralyzed her, urging her to reconsider every step before she made it. She was endlessly second-guessing. The main goal of Sandra's Misguided Protector was to keep her safe, shielding her from potential failure. But this constant catastrophizing of potential scenarios led her to avoid, sabotage, and delay so she never got to test her ideas in the real world. *But hey, at least she was "safe."*

The Ringmaster

Are you the type of person who feels an unstoppable urge to keep pushing forward, to keep working, no matter what? That's the Ringmaster in your head, cracking the whip, shouting, *More! Faster! Better!* It's that voice that keeps telling you: *You should be more productive, You're not earning enough,* or *You can't afford to stop now—there's still so much more to do.* It's what keeps you caught in what Emma Seppälä, a Yale management lecturer and expert in positive leadership, calls the "Never Stop Accomplishing flaw."[7] You hit a goal, and immediately the Ringmaster sets your sights on a new goal. A bigger goal.

The Ringmaster wants to protect you from feeling insignificant. It believes that if you just keep achieving, you'll finally prove your worth. But the finish line always moves. Each win brings a brief burst of satisfaction, only to be overshadowed by a crushing sense of emptiness and that gnawing feeling of "not enough." Psychologists call this phenomenon the "arrival fallacy."[8] It's the lie that says, *When I achieve X, I'll finally be happy.* But instead of the fulfillment you expect, it simply fuels the compulsion to chase the next fleeting high of accomplishment.

The Ringmaster fuels your belief that if you're not always achieving, you're falling behind. And if you stop, you're nothing. As a follow-up to the interviews I was conducting from 2018 to 2020, Fayçal and I conducted a global survey of more than 2,500 people, and a staggering 93 percent admitted to feeling this sort of productivity guilt whenever they stopped working. This is driven by what researchers call "obsessive passion," where your self-worth becomes so entangled with being productive that you don't even know who you are outside of work.[9] Remember our discussion around "Role Fusion" from the previous chapter? This is exactly how it starts.

In the case of Yin, a management consultant, his Ringmaster cracked the whip relentlessly. By his early thirties, he was already a high-flying partner at his firm, but his Ringmaster didn't care. Even after closing the biggest deal of his career, Yin couldn't enjoy the win. *I'm just really passionate about my work,* he told himself, but the truth was that his work had become an obsession, and it was controlling him. His Ringmaster constantly reminded him that he was valuable only as long as he was relentlessly working and achieving. Because of this, he struggled to relax and connect with others outside of work, and any downtime felt like wasted time. He was trapped in an achievement treadmill he couldn't step off.

The Neglecter

The craving for approval and validation is the voice of the Neglecter, that inner whisper that tells you, *You're not enough, for anyone.* This voice makes you feel like you're alone, unworthy, and barely tolerated by the people around you. Usually, this voice takes root early in life. Maybe you didn't get the emotional recognition you needed as a child, and now, as an adult, you're constantly bracing for rejection, convinced it's just a matter of time.

The Neglecter wants to protect you from that rejection by making you chase approval—believing that if you just stay agreeable, you'll never be abandoned. But it comes at a cost: your needs, your voice, and eventually, your sense of self. You bend over backward to earn acceptance, and when it doesn't come, it feels devastating.[10] In close relationships, you might unconsciously repeat these unhealthy patterns. You might find yourself becoming overly dependent on your partner's validation or hyper-sensitive to their mood swings ("Are you mad at me? Did I screw up?"). Setting boundaries becomes nearly impossible, because saying no feels like risking disapproval. And so, you end up overextended, burned out, and—surprise—not feeling good enough anyway. You abandon yourself in hopes that someone else won't.

From early on, Harper's mom was never satisfied with anything she did. As an only child, Harper carried the full weight of her mother's high standards, and the constant disapproval left her feeling invisible and desperate for attention. Now, as a family law attorney, Harper goes above and beyond to please her clients. While that dedication isn't a bad thing, it often comes at the cost of her own well-being. She obsesses over their opinions, fearing their dissatisfaction even when she's given her all. If a client is ever unhappy with an outcome, Harper takes it personally, triggering a deep need to win back their favor: *I have to impress them.* More than their agreement, she craves their validation, convinced her worth depends on it.

What Felix, Sandra, Yin, and Harper needed was to see that these deceivers were wrong. Their logic was flawed. But because the promise of protection felt real, they kept listening. Or more accurately, they didn't know they didn't have to.

Everything begins to shift when you recognize that these voices are simply old patterns on repeat. They're not facts. From that place of awareness, you gain a choice. You don't have to follow the fear, to shrink, or to perform for approval. You can choose not to buy into the lies.

PRACTICE: Standing Up to Your Inner Lies: "Thanks, but No Thanks"

Next time an Inner Deceiver pipes up, especially in a moment when it's hijacking your thoughts and focus, don't get tangled in it. Picture the thought like a bouncing ball flying toward you. Grab your imaginary baseball bat, give it a solid whack, and say, "Thanks, but no thanks." The *thanks* acknowledges that the voice is trying to protect you (even if its logic is completely backward). The *no thanks* makes it clear you're not buying into the lie. And since you're now holding a bat, well, the *protection* part is covered. This taps into the *observer effect*—the psychological principle that the moment you step back and observe your thoughts instead of identifying with them, they lose their grip. This reminds you that you have a choice: to engage, or to let it bounce away. That little pause gives you the space to come back later, with a calmer mind and steadier heart, and start to understand what the voice was really trying to tell you—and why you don't have to believe it.

Talk Back to Your Inner Deceivers

When we understand which Inner Deceiver's voice we hear at any given time, it unlocks something critical—*psychological separation.*

Suddenly you realize there's *you,* and then there's this loud, meddling voice trying to mess with you. And guess what? That voice *isn't you.*

Before understanding their Inner Deceivers, our clients are often stuck in what we call *unconscious assimilation*—absorbing self-criticism and approval-seeking as if they were hardwired traits. They didn't just hear these thoughts—they *became* them. But once they start identifying these voices, they see them for what they are: habitual noise, not truths.

As important as awareness is, it isn't enough when it comes to our Inner Deceivers. We need to give them a literal "talking to." We need to have a heart-to-heart with what that voice is saying, and work through to the other side.

It's what I turned to after my miserable experience speaking to the seasoned executives. I needed to stop spiraling, to take back control of the voices running wild in my head.

So I did something I hadn't done in a long time: I pulled out a spiral-bound notebook, slowed down, and got honest. First, I tuned into the loudest Inner Deceiver: the Classic Judge, tearing apart every part of my talk (right down to my choice of shoe). Then, I reminded myself of what the Judge always forgets: past setbacks that led to growth. I'd faced tough crowds before. I'd stumbled, learned, and adapted.

I even asked myself what I'd say to a client in this exact moment: *One tough audience doesn't define you.*

And finally, I talked back.

Hey, Classic Judge, I appreciate that you're only trying to help me, but respectfully, back off. (The "Thanks, but No Thanks" practice with a bit more thought, intention, and decisiveness.)

Then came the next voice. And the next. I gave each one some space, worked through the same steps, and kept writing.

The forty-five minutes I spent doing this—listening to the

voices and reframing them—didn't change what had happened, but it changed the hold it had over me. I walked away steadier, more grounded, and with a few golden lessons I've carried ever since: *Don't assume who your audience is. Stay true to your voice. Keep moving forward.*

This is how you regain control over the voices that try to derail you. In the moment, emotions might be too charged, but once they settle, come back to this process. Recognize. Remember. Reframe. Respond. And yes, I mean *actually* talking back to those voices in your head.

PRACTICE: Talk Back to Your Inner Deceivers

When you have a bit more space to reflect (outside the heat of the moment), work through the four steps of this deeper practice so you can see your path more clearly:

1. ***Recognize:* Which Inner Deceiver is speaking to you, and what is its real message?**

    The Classic Judge: Fixates on your flaws and mistakes.
    The Misguided Protector: Exaggerates risks to keep you safe but keeps you stuck.
    The Ringmaster: Ties your worth to endless productivity.
    The Neglecter: Seeks validation, making you prioritize others and sacrifice yourself.

2. ***Remember:* When have you challenged this voice before?** Recall a time when you felt the same fear but took action anyway—and it worked out.

3. ***Reframe:* What would you say to a loved one facing this same fear or situation?**

    - If a close friend said, "I'm nervous about this investor pitch," you might remind them, "Of course you're nervous—it means you care. But you know your stuff. Don't worry about being perfect. Showcase your passion and values."
    - If a loved one said, "I'm scared to speak up at work because my boss doesn't like dissent," you might say, "I get it—speaking up when you know it might rock the boat is tough, especially if the stakes feel high. But your ideas have value. Staying silent might be doing your team a disservice." Then, you might encourage them to think about the bigger picture: "What's the worst-case scenario if you share your perspective? If it's constructive and framed respectfully, most bosses, even the tough ones, appreciate when someone brings fresh ideas or identifies problems. If they don't, that says more about their leadership than your worth."

4. ***Respond:* Acknowledge and reassure your Inner Deceiver.** "[Inner Deceiver Name], thanks for caring and trying to help. But I've got this."

Here's how this looked in a completely different context—with my client Vikram.

After being promoted to head of finance, Vikram found himself in a spiral of self-doubt. He started with step one: Which Inner Deceiver was speaking? He quickly spotted his Misguided Protector whispering, *You're not ready for this. You'll fail. Better back out now.*

When I asked him to dig deeper, he paused. "I think it goes back to grad school," he said. "I struggled for the first time and was so embarrassed I shut down for weeks. I guess I'm trying to avoid feeling that way again." That insight was a breakthrough. He saw that his fear wasn't about his job, but an old wound still shaping his thoughts.

Step two: When had Vikram successfully challenged this voice before? He recalled joining a community theater group a few years back: "I've always loved acting, but the idea of being judged or not good enough stopped me from auditioning. Still, I went along with some friends and auditioned, and not only was I accepted, it led to some of the most meaningful friendships of my life." This reminded him that pushing past fear often leads to unexpected rewards.

Step three: What would he tell a friend in his position? After reflecting for a few days, Vikram emailed: "I'd tell them fear is normal, but it doesn't have to stop them. I'd remind them of their strengths, encourage them to take a chance, and say that failure's just part of the process."

By our next session, he was ready for step four: talking back to the Inner Deceiver. He hesitated. "You really want me to say it out loud?" He looked puzzled.

"Absolutely," I said. "Your Inner Deceiver's trying to help you. Acknowledge that—and then take back control."

With some gentle coaxing, he wrote out his response and read it aloud:

"Misguided Protector, I see you. I know you're trying to keep me safe, and I appreciate it. But I've got this. I'm ready to take risks, and I'm okay with making mistakes because that's how I grow."

Vikram was starting to get it—his thoughts didn't have to define him. He was on the road to greater self-acceptance, and with that came a calm he hadn't felt before. He started showing up differently—more present in meetings, less caught in his head, and no longer ruled by self-doubt. He even described feeling lighter, like he'd dropped a weight he didn't know he was carrying.

Of course, his Inner Deceivers still showed up. But this time, Vikram didn't believe everything they said. He challenged them. That shift didn't happen overnight—it came through practice. Through small, repeated choices that slowly rewired his relationship with doubt. And for the first time in too long, Vikram started to trust himself again.

• • • •

Your Inner Deceivers may never disappear entirely, but that's not the goal. The goal is to notice them, question their logic, and choose a better response. Each time you do, you build a new mental habit. Over time, those choices compound. You stop letting self-doubt run the show and start acting from clarity and confidence. That's how real acceptance is built—through the steady repetition of decisions that align with who you want to become.

CHAPTER 6

Why Am I Hiding?

You Are Not Your Masks

In his late teens, Andre Agassi started losing his hair. Embarrassed and terrified of being judged, he resorted to wearing a hairpiece to conceal it. The breaking point came shortly afterward, during the 1990 French Open, where he reached his first Grand Slam final.

"The night before the final," Agassi recalled, "catastrophe strikes."[1] After he used the wrong conditioner, his hairpiece began to fall apart. In a panic, he salvaged it with twenty bobby pins. Instead of focusing on the biggest match of his life, Agassi was consumed by anxiety about his hair. "I prayed," he later confessed, "not for a win, but for my hairpiece to stay on." That fear of being exposed, of others noticing his vulnerability, was so overwhelming that it completely threw him off.

He lost the match—three sets to one.

Agassi's insecurity about what others would think led him to sacrifice his performance in one of the biggest moments of his life. And many of us do this too. Maybe not with hairpieces, but we all wear masks to hide parts of ourselves, hoping to win approval or dodge judgment.

In my case, I'm terrified of hurting other people's feelings. I'm the kind of person who buys something I don't want because I feel bad for taking up the sales assistant's time. I've been known to quietly sit and eat a meal I didn't order just to avoid inconveniencing a waiter. I've smiled and nodded through conversations that didn't sit right. Laughed at jokes I didn't find funny. Found a thousand ways to agree with people to avoid the sting of disconnection.

For as long as I can remember, I was good at doing what I thought was necessary to be liked and accepted. Maybe it came from being the daughter of immigrants. Maybe it was a result of early experiences I had at school. Maybe it was how I adapted to my parents' divorce. Or maybe it started with my personality. Whatever the reason, I learned to be *agreeable*—and not just in a pleasant way, but in a way that became habitual. A habit that became a constant mask.

The Danger of Your Masks

Agreeableness in itself isn't a bad thing. In fact, when I asked our online community to list the qualities associated with agreeableness, here's what they shared:

Empathy. Generosity. Unconditional kindness. Selflessness.

The research agrees.[2]

Then I asked them to share the qualities of a *people pleaser*. They shared a very similar list, but with the addition of these more challenging qualities: *Difficulty saying no. Pretending to agree when you don't. Over-apologizing. Trying to "fit in." Desperate for validation.*

That right there is the difference. Agreeableness is healthy. People-pleasing is agreeableness on steroids. It's lost all semblance of moderation. Social psychologists call it *sociotropy*.[3] It's when our desire to be accepted becomes so strong that we start shrinking, contorting, and second-guessing ourselves—just to avoid judgment, rejection, or conflict.

If you're an approval-seeking people pleaser (or, like me, a recovering one), you know what I'm talking about. You care way too much about what others think. And no matter how hard you try to ignore it, that little voice creeps in: *Is that driver judging the way I look while I jog? Does that client see me as a screw-up because of that typo? Is my latest post cringeworthy?*

And slowly, almost imperceptibly, you begin to compromise. You water yourself down, sacrificing what you actually want, need, or believe—just to keep everyone else happy.

During my PhD research in 2022, I conducted a series of anonymous surveys to explore the relationship between self-criticism and work performance. After the first survey, one participant emailed me to say, "I gave answers I felt would make you proud." (And yes, this has been raised as a limitation of the study.)

The irony is, the survey was anonymous, and I emphasized the importance of candid responses. But still, the person felt the need to "impress" me—a total stranger—by presenting a version of themselves they thought I would approve of.

The term for this is *social desirability bias*.[4] It's when people hide their true opinions or "fake" their responses to make themselves look better to others. In other words, they reach for a mask.

Whether it's in an anonymous survey or in our day-to-day lives, it shows just how powerful the desire for approval—and thus acceptance—can be. We tailor a version of ourselves that we believe is "acceptable" to others, which means that by default, we believe *we* aren't acceptable as we are. The entrepreneur and author Mo Gawdat explains it this way: "We wear different masks and hide our reality from everyone, including ourselves. Our assumed identity becomes our whole lives, and we start to believe them—even more than others do."[5]

What starts as a social habit becomes a second skin.

Psychotherapist Phil Stutz describes what happens when you begin to hide parts of yourself. He says it doesn't end with the hiding. Instead, we become hyper-attuned to those very parts we want to cover up, forever scanning to see if others might see them too. "It becomes an obsession," he writes. "How do they see me, what do they think of me, do they like me, love me?"[6]

To protect ourselves from that obsessive focus on our flaws, we reach for masks we hope will hide them.

Sometimes the mask is obvious, like dressing a certain way (or wearing a hairpiece like Agassi) to fit in. Other times, it's subtler: always trying to give the "perfect" answer, apologizing for no reason, forcing a smile, saying yes when everything in us wants us to say no.

I know this because I've lived it. Looking back, I see how it showed up in the early years of my corporate career. At the time, I didn't feel like I was *enough* as I was, so I needed to "fake it"—to speak a certain way to sound more credible. To use complex words to sound more intelligent. To wear the right clothes and maintain a perfectly composed (almost cold) demeanor. I felt like I had to constantly put on a show to be taken seriously. I was *performing*.

Even now, all these years later, I still catch myself slipping into the corporate mask I wore for so long. Before a big meeting I have to remind myself to "loosen up," or before a client call I'll tell myself to "drop the formal corporate voice." I'm still unlearning the habit of masking.

So why is wearing masks so dangerous? Because the moment you summon the courage to step out of that mask and show up as your real self—to assert yourself, share your opinions, or do what you *actually* want—any flicker of disapproval can feel like a threat. Your instinct is then to scramble to win back the lost approval. You put the mask back on. You stop trusting yourself. You doubt your instincts. Slowly, you retreat to your familiar patterns of hiding, doubting, and chasing validation.

Every time you hide a piece of yourself, you give that hidden part more power. You're telling yourself, *This part of me isn't good enough to show the world.* You reinforce the belief that your worth lives in someone else's eyes. That's how self-doubt takes root and flourishes.

The Courage to Drop the Pretense

Let's pause for a moment and consider this: Is wanting others to accept you and think highly of you inherently wrong? Not necessarily. It's human. We all want to feel seen and valued—especially by the communities we care about. When someone recognizes your work or compliments your character, it lights up the reward pathways in your brain.[7] It feels great. It reinforces belonging. If you respect someone's skills, wisdom, or way of being in the world, it's natural to want their validation. That's how we learn and grow.

It's also a beautiful thing to want others to be happy when our efforts come from a place of love, care, or generosity. The problem arises when these actions stop being a choice and become a compulsion. People-pleasing drains you because it comes from fear, not genuine desire. On the other hand, doing things for others from a place of true care and choice means giving without sacrificing your own integrity or energy. It's a world of difference, and it feels entirely different too.

When your self-worth *depends* on others' opinions, you lose the ability to accept yourself. You never feel like you're "enough" as you

are. As Ronda Rousey, former UFC women's bantamweight champion, so aptly puts it, "Once you give them the power to tell you you're great, you've also given them the power to tell you you're unworthy. Once you start caring about people's opinions of you, you give up control."[8]

We're constantly bombarded with messages about what a "successful" life "should" look like: how much money we *should* make, the career we *should* have, how we *should* look, dress, date, parent, and even spend our free time. It's exhausting.

If you manage to meet society's "shoulds" and standards, you might gain its approval, and maybe even a sense of satisfaction—at least for a while. But what happens when you fall short? Or when you realize the goal itself doesn't feel meaningful? That's when the self-doubt creeps in. *What's wrong with me? Why can't I be good enough?*

When you're chasing standards that were never yours to begin with, you lose sight of your own values, desires, and identity. And without those? No mask in the world can hide how hollow that feels.

PRACTICE: "Stop Should-ing Yourself"—Peel Off a Mask for a Day

Masks often come disguised as "shoulds."[9] Pick one *should* that's been draining you—something you do for approval rather than for yourself. Now, peel that mask off, just for one day. Notice what happens when you release the rule, even briefly. For example:

- If you believe **you *should* always have the answers,** ask for advice instead.
- If you feel **you *should* filter your opinions,** respectfully share your honest perspective.
- If you think **you *should* dress a certain way to fit in,** wear something that reflects *you*.
- If you believe **you *should* always be available,** mute your notifications tonight.

Now it's your turn: *If I believe I should* ________, *then it's time to* ________ *instead.*

If a coworker raises an eyebrow at your bluntness, or a friend teases you about your bold outfit, that's okay. The goal isn't to keep everyone else comfortable; it's to see how you feel without the mask.

The Freedom of Knowing and Being Yourself

After his crushing French Open loss, Agassi's then-girlfriend Brooke Shields gently suggested he ditch the hairpiece. "Impossible," he said. "I'd feel naked . . . exposed." But she saw it differently: "You'd feel liberated."

At first, he resisted. Stepping onto the court without his signature lion's mane felt unbearable. But eventually, he gave in. He shaved his head and, in doing so, shed his disguise.

When Agassi showed up at the 1995 Australian Open, bald and unburdened, he dominated. In the final, he defeated defending champion Pete Sampras, three sets to one. Agassi later admitted, "My hairpiece was a shackle."

In his memoir *Open* (which, fittingly, features a close-up of his now-bald head), he wrote: "Everyone says it was my best performance yet . . . but I think twenty years from now I'll remember this as my first bald victory." Quite literally, he peeled off a layer of self-doubt that had been stuck to him. He tapped into that place of Big Trust—that larger sense that he was worthy and could accept himself just as he was.

In my own life, growing up, I learned to put others first, to never be a burden. My brother, who looked visibly Middle Eastern, faced discrimination I didn't, especially after 9/11. I saw how much energy my parents poured into supporting him, so I tried to be easy—the low-maintenance child who could manage everything on her own. My way of showing love was to *not* need anything at all. That role I adopted became part of my identity.

That mask followed me into my teenage years. I became the good kid, and then the head prefect at school (Australia's version of student body president), polished on the outside but twisting myself into knots on the inside.

When I started working, I wanted to be liked, so I made myself *helpful.* I volunteered for every thankless task, especially taking minutes in meetings. Every. Single. Meeting.

Initially I felt like a team player. Yet the more I said yes to those small things, the more I said no to being fully seen. Over time, instead of feeling valued, I felt invisible.

It wasn't until I joined a newly formed leadership team that I saw how deep the habit ran. In our first meeting, I felt that old urge to raise my hand and offer to take notes, but I knew I couldn't keep doing this to myself. I literally sat on my hands to stop myself. It felt ridiculous, but it worked.

The group decided to rotate note-taking. For the first time, my role was to contribute ideas and challenge opinions, not just transcribe them. I actually felt like a leader. The difference was night and day, not just in how others saw me, but in how *I* saw myself.

That so-called "helpful" mask hadn't helped me at all. It had muted me. It had kept me small.

Unlearning the belief that my worth was tied to making others happy took practice. I stopped apologizing for things that weren't mine to carry. I stopped saying yes out of habit or guilt. I started speaking up, even when my perspective differed from the group.

That sense of self-trust over self-doubt was tested again and again when I started posting online.

There I was, this recovering people pleaser, suddenly exposed to hundreds of thousands of opinions. Some loved the content. Others . . . not so much. They nitpicked my appearance, pointed out flaws I hadn't even noticed, and criticized everything from my uneven eyebrows (so I fixed them) to the way half my face twitches when I speak (nothing I can do about that).

This was the ultimate practice in *letting go of what others think*.

One of the hardest truths to accept when you struggle with acceptance is that not everyone will like you (you're not pizza!). I once shared the pizza example with an audience, and a guy yelled, "I don't even like pizza!" Case in point: You really can't win them all. And that's okay. Someone doesn't like you? That's their story. Someone judges you? They're just revealing their insecurities. Someone disagrees with you? That's allowed. As bestselling author Mel Robbins would say, "Let them."

With every layer of self-doubt I let go of, and every mask I peeled away, I felt a little more free. In my own small way, each time I let go, it felt a little like Agassi's first bald victory. Embracing who I am—face twitch and all—has been one of the most liberating steps in my journey.

Freedom doesn't erase the temptation to perform. It might still show up, but now you'll notice it. And when you feel the familiar pull to reach for the mask again—to bend, appease, make someone else comfortable at the cost of your own peace—pause. Take a breath. Researchers from Columbia University have found that even a tiny pause (of a fraction of a second) can help us make better decisions under pressure.[10] You can even sit on your hands if you need to, like I did to keep myself from volunteering for a note-taking role I didn't really want.

Ask yourself, "Does this response align with the version of me I want to become?" As author Stephen Covey reminds us, "It's easy to say 'no' when there's a deeper 'yes' burning inside."

Say *no* to extra commitments for a *deeper yes* to finally focusing on what you've been putting off.

Say *no* to working late for a *deeper yes* to dinner with your kids.

Say *no* to a toxic relationship for a *deeper yes* to self-respect and emotional safety.

Say *no* to pretending you're fine for a *deeper yes* to finally asking for support.

Say *no* to your family's dream for your life for a *deeper yes* to your own dream.

Say *no* to being "the reliable one" for a *deeper yes* to being the *real one.*

Every pause creates space. Every no creates the possibility. And in that space, you give yourself permission to drop the mask and accept who you are and who you want to become.

Because if you spend your whole life performing, there will come a day when the show ends—and you'll wonder what it all was for.

Bronnie Ware, a palliative care nurse, spent years sitting beside people in their final days. She wasn't a researcher or a psychologist, just someone who was there when the masks had come off, when the roles had fallen away, and all that remained was truth. In her book *The Top Five Regrets of the Dying,*[11] she shares the most common reflection she heard. It wasn't "I wish I'd been more liked," or "I wish I'd said yes more often." It was this: "I wish I'd had the courage to live a life true to myself, not the life others expected of me."

PRACTICE: Align with the Person Behind the Mask

This simple four-step practice[12] will help you recognize approval-seeking, challenge the fear behind it, and realign with your true self.

Step 1: Ask, "Why Am I Seeking Approval Right Now?"

Get brutally honest with yourself. What's driving this urge to please?

- Wanting to be liked?
- Searching for belonging?
- Doubting your value?
- Afraid of rejection or conflict?
- Hoping to control an outcome?

Pinpointing the *real* reason can reveal deeper insecurities at play.

Step 2: Ask, "What Is My Need for Approval Costing Me?"

Approval-seeking isn't free. It costs you something. Maybe it's pulling you away from your values, making you compromise your boundaries, or silencing your voice. Remember, every yes rooted in fear is a quiet no to something that matters.

Step 3: Ask, "What If I Acted Like I Already Had Approval?"

Imagine if you already had the approval you're seeking. How would you show up differently? When you self-validate instead of outsourcing your worth, you naturally project more confidence, and ironically, that's what earns genuine respect.

Here's what this might look like:

- **In a meeting:** Instead of editing your ideas to be more "palatable," you share your perspective clearly, without softening the edges.
- **At work:** When your supervisor compliments you, you don't deflect with "Oh, it wasn't a big deal." You simply say, "Thank you. I put a lot into it."
- **With friends:** Instead of tagging along with the group's plans to keep the peace, you suggest something you actually want, like "I'd love a quiet night instead."
- **With family:** Instead of nodding along to avoid disagreement, you kindly speak your truth, like "I see it differently, and here are the reasons."

Step 4: Ask, "What Will I Regret *Not* Doing?"

Zoom out. Picture yourself at the end of your life, looking back.

- What will matter most?
- What would you wish you'd said, done, or become?
- What risks will you wish you'd taken?

Now bring that clarity into the present moment:

- If today were your last day, what would you do differently? Who would you *choose* to be?
- And what's one small way you can start living in alignment with that, today?

The more often you choose what matters over what pleases, and the more you begin living in alignment with your values, your voice, and the version of yourself behind the mask . . . the more you will become the *you* that doesn't need to perform to feel worthy.

• • • •

For so long, you've worn a mask, reshaping yourself to fit expectations, and to furiously quiet the fear that you're not enough. But what if you let it slip, even just a little? What if you stopped bending to fit in and started standing in what actually matters to *you*?

Instead of performing, practice the habit of honesty—not just with others, but with yourself. Start small. Tell the truth about what you need. About how you feel. Practice honesty, even if your voice shakes. Little by little, you'll realize you were never waiting for permission or approval. You were waiting to trust yourself.

CHAPTER 7

Do I Have to Be Perfect?

The Fallacy of Perfectionism

During the COVID-19 lockdowns, the world turned to social media not just for an entertaining diversion, but for support, insight, and comfort. Fayçal and I certainly needed that. Like so many millions of others, COVID-19 derailed our plans. I'd just wrapped up a decade of corporate work, we were about to move from Australia to Southeast Asia, and we were launching our own business. Then the world stopped.

For months, we sat in the limbo of *What now?* The urge to press pause until the path felt clearer was strong. But I knew that if I waited until everything was perfectly figured out, I'd never take the leap to start building the kind of work I truly wanted to do.

Fayçal knew that too. "You don't need a perfect plan to start helping people," he'd tell me. "Who cares if it's unconventional as long as it's useful?"

I'd nod my head—and then do nothing. However, on the inside, my Risk Analyst was working overtime, asking me all the questions that often stopped me from starting big projects:

What if it's not good enough?

What if people judge me?

What if I fail publicly?

Now, my idea was simple. Through video on social media, I knew I could help the overwhelmed, the anxious, and the uncertain—the millions navigating job losses and identity shifts. (Fayçal thought TikTok was for sixteen-year-olds, but he backed the idea because he saw how fired up I was to serve.) While I didn't have all the answers, I could offer the tools I'd gathered through my MBA in innovation and leadership, through studying positive psychology, and through becoming a certified executive coach. Tools for resilience. Tools for clarity. Tools for navigating the inner chaos that often follows outer change.

But I'd have to outsmart my perfectionist first.

So I asked myself, *What if I made short, practical videos?* Bite-sized lessons people could apply right away. I mapped out forty topics. (Why forty? It just felt right.) I scribbled bullet-point scripts and filmed them all in a single day—really. My goal wasn't to make them brilliant or to hope for thousands of views. I just needed proof that *done* was better than *perfect*. My setup wasn't fancy: just my phone, a basic tripod, and a few quick changes of jackets, earrings, and hair to keep things fresh.

The filming itself wasn't particularly difficult. It was what came next. The moment of truth. The part where I *actually* had to post them. I sat there, phone in hand, staring at the screen. My thumb hovered over the "post" button, but I couldn't press it.

The fear was instant, familiar, and loud. Full-body resistance. My heart pounded. My stomach turned. Every instinct said, *Wait. Tweak the caption. Fix the lighting. Maybe re-record them all.* Anything to delay the risk of putting myself out there. Because once I hit post, there was no going back. The longer I sat there, the louder the self-judgment became.

And in that moment, I caught myself.

This was the exact trap perfectionism had set for me before. The endless waiting, preparation, and planning. The illusion that if I could just tweak and polish enough, I'd finally feel ready.

I wasn't ready. But I posted anyway. One video on a Monday morning. Then another. And another. Unpolished. Nowhere near perfect. But done. Each time, the fear surged. And each time, I posted anyway.

For almost three weeks, nothing happened. Barely any traction. I had every excuse to delete them, to tell myself, *See? It's not working.* But I didn't. Because I wasn't measuring success by views or likes. I was measuring it by the act of posting. I told myself, *Even if just one person finds calm or courage through these videos, it's worth it.*

And then, on day twenty-one, one video on the science of making a lasting first impression went viral. Tens of thousands of views in a matter of hours. Then hundreds of thousands.

Four weeks in, we went from zero to 75,000 followers. A couple of weeks later, 225,000. Media outlets started paying attention.

All while in lockdown, we grew a community of nearly 2 million people, were featured in major media outlets, and were invited to work with Fortune 500 clients all over the world. A few years later, those videos we filmed in our humble Melbourne apartment had been viewed more than 300 million times and our audience grew to more than 5 million.

And it all started with forty imperfect videos.

The Problem with Perfectionism

How many moments have you delayed, not because you didn't care, but because you cared *so much* it felt safer to wait? What part of your life is waiting for you to let go of "perfect"—and just begin?

Perfectionism likes to pretend it's about high standards, but that's a lie. It's not about excellence—it's about safety. It's the inner voice of your Misguided Protector whispering, *Don't risk it. Stay small. If you never put yourself out there, no one can reject you.* It tricks you into thinking that if you get everything just right, you can avoid embarrassment, rejection, or being seen as *less than.*

Researcher and leadership expert Brené Brown calls perfectionism "a twenty-ton shield" that we carry, hoping it will protect us from judgment.[1] In *Atlas of the Heart,* she describes it as a dangerous and debilitating belief system that tells us, "I am what I accomplish and how well I accomplish it. Please. Perform. Perfect."[2] It's a moving target, an unwinnable game where you measure your worth by an ever-elusive ideal. You strive for "flawlessness"[3] and then berate yourself when, inevitably, you don't hit the mark.

The irony is that while you're chasing "flawlessness," you're not living. Perfectionism stops you from trying new things, finishing projects, or taking risks. The fear of failure, or falling short, convinces you that other people's opinions have the power to destroy you. But often, this fear itself leads to the very failures you're trying so hard to avoid. Because you wait. You hesitate. You tell yourself, *I'll put myself out there when I'm sure it's bulletproof.* You stay trapped in the illusion that "more effort" or "more planning" will bring perfection closer.

But in trying so hard to avoid rejection, you reject *yourself* first.

If you wrestle with acceptance, your failures—or perceived failures—feel personal. It becomes an identity: *I failed, therefore I AM a failure.* To protect yourself from feeling that again, you tend to go one of two ways: Either you avoid risks altogether, or, if you're like most high achievers, you swing to the other extreme. You chase perfection. You raise the bar higher with every goal, every project, and every relationship. The idea is: If I make no mistakes, no one can criticize me. But that strategy doesn't add to your success—it blocks it.[4]

The only way out is to rewrite the belief and break the perfectionist habit that feeds it. Like I did with my forty videos in forty days, you have to let go of "flawless," and start choosing "finished." Otherwise, you'll keep measuring your worth against an impossible standard, and always come up short.

Perfectionism works against you in another way—it multiplies. You don't just set unattainable standards for yourself; you project those standards onto others too. You start worrying about not only what you're doing, but also how others are judging it. You double, triple, or quadruple the potential for even more negative judgment and self-doubt.

The truth is, most people don't think about you nearly as much as you think they do. Known as the *overblown implications effect,*[5] we tend to wildly overestimate how much others care about us and our perceived failures.[6]

But of course, even when we know something intellectually, it can still affect us. When we worry about being judged or criticized by others, we start to internalize those fears and obsess over them, which only magnifies the feelings of humiliation and shame—even before anything has actually gone wrong.[7]

Social scientists at UCLA conducted an illuminating experiment some years back that showed how our brains respond to social pain, like rejection.[8] The results revealed that the same brain regions that light up when we experience physical pain also activate when we face social rejection. Yep, getting criticized or judged activates the same pain centers as burning your hand on a hot pan or stubbing your toe. That's why we're so desperate to avoid it. And we can. We can gradually unhook from perfectionism and the sting of perceived criticism, like pulling burrs from Bonbon's fur and my socks after a walk.

"Good" Perfectionism Versus "Bad" Perfectionism

Let's face it, falling flat on your face is basically life's favorite way of teaching you something. It's the universe's way of handing out free lessons—painful, embarrassing, but undeniably effective. Failing isn't just part of the process; in most cases, it *is* the process.

When James Dyson, the man behind the billion-dollar vacuum empire bearing his surname, was in the early days of developing vacuum prototypes, he struggled. "By the time I made my 15th prototype, my third child was born. By 2,627, my wife and I were really counting our pennies." Looking back, Dyson sees that this struggle was just part of the invention process. He shares that it was each "failure" that brought him one step closer to solving the problem. "It wasn't the final prototype that made the struggle worth it. The process bore the fruit. I just kept at it."[9] After 5,127 prototypes and fifteen years, he finally got it right.

For Dyson—or any inventor, artist, professional, entrepreneur, or leader—the real danger isn't just the financial strain or lost time. The real risk is when failure feels like confirmation that *they* are the problem. Not the prototype. Not the process. *Them*. And that's where things get dangerous. If you struggle with self-acceptance, you don't just see the failure, you *become* the failure. And from there, it's easy to spiral into unhealthy perfectionism, where every mistake feels like it chips away at your worth.

When Fayçal and I run our workshops, we'll often find someone asking a question along the lines of, "I set high standards for myself—is that perfectionism?"

The answer is: "It depends."

It boils down to *why*. Why are you setting those standards? What's driving you? If your goal is to do your best, to learn, and to grow, that's not perfectionism—it's what we call "healthy striving for excellence." Researchers call it "excellencism."[10] It's the mindset that asks, "How can I do better next time?" instead of "What will they think if I fail?" When you aim for excellence, you're not afraid to mess up because you see mistakes as part of the process. We also know from research that if you focus on growth and learning, you're more likely to achieve your goals and have positive life outcomes.[11]

Unhealthy perfectionism, though, is a different beast. It doesn't drive you—it traps you. It's fueled by fear, by the nagging belief that one misstep will shatter the carefully crafted image you've built. The hyperactivity and constant push to meet sky-high standards can put you on constant high alert, turning you into someone who's always on edge, hypervigilant, and overly harsh on yourself for not hitting your impossible targets.[12]

One of my clients summed it up perfectly: "My perfectionistic tendencies drain my world of color, light, and spontaneity. It feels like I'm in a prison, 'pretending' to live a perfect life, but I'm suffering."

How to Prevail over Perfectionism

Elizabeth Gilbert's struggle while writing her memoir *Eat, Pray, Love* is the kind of perfectionist nightmare that anyone who's ever tried to create something can relate to. Self-doubt clung to her every move, making her question every sentence, every page, every paragraph. She recalled, "I had just as a strong a mantra of THIS SUCKS ringing through my head."[13] Every page felt like a failure waiting to happen.

But amid this storm of self-criticism, Gilbert found a moment of clarity: "The point I realized was this—I never promised the universe that I would write brilliantly; I only promised the universe that I would write. So I put my head down and sweated through it, as per my vows." Instead of chasing perfection, she made a decision: Focus on the process, not the outcome. Forget "write brilliantly." Just write. That mindset shift didn't make the fear vanish, but it let her move forward despite it.

The lesson here is clear: You won't always feel like you've produced your best work. Sometimes it won't feel as polished as you'd like. But letting that hold you back is often far worse than putting something out there that might not meet your own high standards.

Like when I sat there staring at the "post" button, feeling that full-body resistance, convinced I should wait—until the lighting was better, the videos were sharper, until I felt "ready." If I had given in to that urge, those forty videos would still be sitting on my phone, unseen. Instead, they became the foundation of something far bigger than I ever could have imagined.

Don't let "perfect" get in the way of "good enough for now." You can always polish later. But if you wait for perfection, you might never hit publish, launch, or step on that stage.

Plus, you never know the impact that your version of "good enough for now" might have.

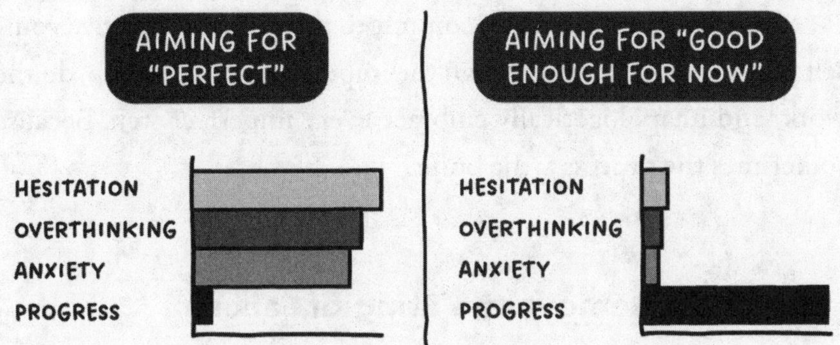

PRACTICE: Process over Perfection

An important part of moving past perfectionism is to make a vow to focus on the *process* rather than obsessing over the *outcome*. Take a page out of Elizabeth Gilbert's playbook and just promise yourself to "write," not "write brilliantly." The key is to increase your tolerance for imperfection so that you don't just tolerate it but learn to embrace it.

Marketing expert Seth Godin has a term for this: "meeting spec" (where *spec* stands for *specification*).[14] Translation? Just get started with the minimum. Don't worry about flawless for the vast majority of your efforts; aim for functional. Make the simplest action the goal.

Here's how this looks:

- Don't aim for "find my soulmate"; just "ask them out for a coffee."
- Don't stress about "going viral"; just post something on LinkedIn.
- Don't wait until you have the perfect product; just test an idea with real people.
- Don't make "succeed" the objective; just start.

Magic happens when you commit to the act itself and give yourself permission to grow through the imperfections. Show up, do the work, and unapologetically embrace every imperfect step. Because sometimes the process *is* the point.

Remove the Sting of Failure

The reality is, perfection is a unicorn. You'll never catch it. And even if you could write the perfect email, pitch the perfect deal, or deliver the perfect presentation, there's no guarantee it'll land the way you want it to. Just ask James Dyson. By prototype 2,627, he still had no assurance his vacuum would ever work. The risk of "failure" is always there, lurking in the background.

I learned early in my career that it's possible to take the sting out of rejection and failure. I call it "Quack Your Duck." For a brief, soul-sucking period, I worked at an inbound call center where 70 percent of calls were from very, *very* unhappy customers. It was tough. The barrage of negativity we received every day took a huge toll on morale.

The management team came up with a quirky solution. They gave each of us a small plastic duck. Squeeze it, and it would let out a comical "quack!" sound. The idea was simple: Get a grumpy caller, end the call, squeeze your duck for the whole floor to hear, and then tally up your "unhappy customer" count on the whiteboard. It was a stroke of genius, really. They turned a stressor into a scoreboard, and just like that, the dread started to lift. There was something oddly satisfying about marking those tough calls and hearing the chorus of quacks. We began to anticipate the difficult calls, not with fear, but with a readiness to turn them into points. It took the edge off, and those negative calls didn't feel so personal anymore.

Even if you're not fielding customer service calls all day, you can steal the "Quack Your Duck" mindset for those moments when perfectionism or failure is clawing at you. Your duck doesn't have to be a literal quack machine. It's simply a reminder that you don't have to take rejection or criticism so personally—that you can step back and soften the sting of those moments. Whether you're navigating challenges at work, in your personal life, or on a creative project, creating your own version of this ritual can be incredibly helpful. It might be as simple as a deep breath, a quick walk, or having a small, lighthearted habit that helps you shake off the negative emotions. Maybe a feel-good playlist or a quick stretch. The key is to stop taking rejection and criticism as proof that you're broken. Acknowledge the moment, then "Quack Your Duck" and move past it.

PRACTICE: Honor Your Failures

Many high achievers believe they need to be tough on themselves to stay competitive, motivated, and on top of their game. They think relentless self-criticism is what keeps them pushing forward. But it's not. That's just their Ringmaster talking trash in their head. The research says otherwise. People who are more accepting of their so-called shortcomings and "imperfections" are actually better at growing, improving, and, yes, succeeding.[15]

So take a good, hard look at your approach. Be real: Is it really working for you, or is all that self-judgment getting in your way? It's time for a Reality Check, and then for you to extract the lessons from the experience to give it meaning. This process is similar to what's called a postmortem—a practice often used in project management to analyze what went wrong, what went right, and what can be learned for the future.

1. **Name Your Judgy Thoughts:** Think back to the last time you made a mistake, "failed" at something, or just felt like a walking disaster. What were your Inner Deceivers whispering to you at that time? Write it down.
2. **Conduct a Reality Check:** Look at the situation objectively. No harsh self-talk, no blame. What really happened? Be honest. Stick to the facts.
3. **Create an Excellence Statement:** Excellence doesn't require you to get everything right the first time. It's about using *what went wrong* as a launchpad to *get better*. Consider:

    - What did you learn from this experience?
    - What do you know now that you didn't know before?
    - What small shift could help you achieve a better outcome next time?

This becomes your Excellence Statement. It's your way of "Quacking Your Duck," by calling out the "imperfection" or "mistake," reframing it, and increasing your odds of success next time. You're reworking your thoughts—what psychologists call *cognitive restructuring*—to help you move forward without dragging the weight of self-judgment behind you.

Here's how this might look in practice:

Situation: You turned up to a client meeting unprepared.

1. **Your Judgy Thoughts:** "I'm incompetent and don't deserve this role."
2. **Reality Check:** "I showed up unprepared, which meant I couldn't contribute as much as I wanted to."
3. **Excellence Statement:** "I'll learn from this. Next time, I'll

block out thirty minutes the day before to review the brief and jot down key points I want to address."

Situation: You missed crucial plays in a basketball game.

1. **Your Judgy Thoughts:** "I'm worthless as a player and don't belong on the team."
2. **Reality Check:** "I was so focused on the other team's strengths that I psyched myself out and lost confidence."
3. **Excellence Statement:** "I'll improve my focus and build my confidence through practice and running mental drills. If I notice myself getting distracted by the other team, I'll consciously refocus on what I can control: my moves, my strategy."

Situation: You made an error in a report.

1. **Your Judgy Thoughts:** "I messed up, and now everyone thinks I'm careless."
2. **Reality Check:** "I missed a key detail in the report, which caused some confusion and needed correction."
3. **Excellence Statement:** "I'll learn from this. From now on, I'll proofread my work twice to make sure it's more accurate. If it's a high-stakes report, I'll ask a coworker to double-check critical sections with me."

Now, if your mind is anything like mine, you might find your brain will fight you on this. One minute you're cooly conducting your Reality Check, and the next, you're spiraling into "I'm the worst human alive." The key is to find the right environment to stay in the Reality Check mode. For instance, I've realized that watching a recording of a keynote I've just delivered is a no-go for me. I'm too quick to criticize myself. If it's too soon after the event, I can't separate my immediate emotional responses from a more balanced evaluation. Knowing this about myself, I have a rule: No watching replays until at least twenty-four hours have passed. This brief distance helps me return with a clearer, more compassionate, and more objective perspective. Then I ask myself, "What can I tweak to make the next one even better?"

Try flipping the script on how you view errors and setbacks. Recognize them for what they are—a chance to learn, adjust, and come back stronger. Perfectionism warps our perception of failure, making us believe that every mistake is a verdict on our worth. But in reality, some of the biggest breakthroughs come from missteps. (Just think, we wouldn't have penicillin if not for Scottish scientist Alexander Fleming's "mistake" where he accidentally dropped a mold spore into a petri dish of bacteria.)

Failure can be your greatest teacher, but only if you take the time to reflect and extract the gems hiding in the rubble. Honoring your losses helps you detach your self-worth from the result and see the bigger picture. Over time, this kind of reflection becomes a habit, one that trains your brain to look for growth instead of shame.

• • • •

Perfectionism isn't only driven by a fear of failing or falling short—it makes you afraid of being *seen* as failing or falling short. It's not the mistake itself that keeps you up at night, but the shame story that follows: *What will they think of me? What does this say about who I am?*

But failure is part of life. And perfection is an elusive target. The more you chase it, the further away acceptance feels. Having self-trust doesn't mean you'll never stumble. It means you refuse to let perfectionism turn the possibility of failure into something that stops you.

CHAPTER 8

The Gift of Self-Forgetting

Acceptance in the Here and Now

Finding Acceptance in the messy, real world of work and life is a constant work in progress. On the one hand, you can peel back the beliefs and worries you've grooved into your mind and the inner voices, stories, and perfectionism that undermine you. And on the other hand, you can turn to and build on the strengths you already have. But there is a third way to experience Acceptance—a way that helps propel you beyond any struggle or search for worthiness. You can set aside all the tools and insights and step into something bigger than yourself. It's a principle taught by 'Abdu'l-Bahá, the spiritual teacher and son of the founder of the Bahá'í Faith: "The key to self-mastery is self-forgetting."[1]

In 2010, Fayçal was hired to coach and write a speech for the host and moderator of a G20 Summit. This wasn't just any public speaker. He was a billionaire, the chairman of one of Asia's largest conglomerates—a man well-versed in power, influence, and high-stakes decision-making. And yet, in the lead-up to the event, he was quietly unraveling.

His English wasn't fluent. His accent was heavy. And all he could fixate on was the fear of underperforming and being judged. He wasn't thinking about the speech. He was trapped in his own mind, caught in a spiral of self-doubt.

Fayçal developed a speech for him—concise, powerful, easy to deliver. But no matter how much they rehearsed, the doubt didn't budge.

With only days to go, Fayçal realized that encouragement and practice alone wouldn't cut it. This wasn't about delivery technique. It wasn't a skill issue. It was a mindset issue. And they needed a breakthrough—*fast*.

"Then it hit me," Fayçal later told me. "I needed to help him shift his perspective. I said, 'You're not just here to formally present to these leaders while worrying about your accent being scrutinized. This isn't about *you*. It's bigger than that. You're here to set a tone of fellowship and unity—so these world leaders feel empowered to work together to improve the well-being and prosperity of humanity, while protecting our planet. When you're on that stage, focus on *that*—and the inner chatter will quiet down.'"

It didn't take long for the shift to take effect. The chairman's posture relaxed. His voice steadied. His demeanor became more confident. The weight visibly lifted from his shoulders as he shifted from focusing on *himself* to focusing on the *mission*.

The problem was never his accent. It was that he had made himself the center of his attention. And the moment he shifted the focus outward, his confidence followed.

I've seen this transformation over and over—in keynote speakers, in CEOs preparing investor pitches. In everyday people stuck in their heads, convinced they don't belong.

And here's the pattern: The ones who struggle the most are the ones who can't stop monitoring themselves. They are caught in a feedback loop of self-consciousness, where every move is measured, every flaw magnified, every word second-guessed.

And the irony? They could be brilliant. They could be prepared. But none of that matters if their attention is trapped in self-doubt because they don't accept themselves.

Viktor Frankl understood this too. The Austrian neurologist, psychologist, and Holocaust survivor penned dozens of books. But the one that changed the world?

It was never meant to be a success.

His seminal work, *Man's Search for Meaning,* was originally published anonymously. Frankl never intended for it to become widely known, let alone a bestseller.

Throughout the rest of his life, he shared the lesson he'd learned from the book's unexpected reception: "Don't aim at success—the more you aim at it and make it a target, the more you are going to miss it. For success, like happiness, cannot be pursued; it must ensue, and it only does so as the unintended side-effect of one's dedication to a cause greater than oneself." "In the long-run," he says, "success will follow you precisely because you had *forgotten* to think about it."[2]

Remembering "Self-Forgetting"

This is the essence of self-forgetting:
Redirect attention away from yourself—your insecurities, your ego, your self-consciousness—and toward something bigger.

I learned the power of self-forgetting firsthand early in my banking career.

In my second year, I applied for an opportunity to spend six weeks living with a remote Indigenous community. That's how I found myself in Kowanyama in Cape York, Far North Queensland, Australia. I didn't fully grasp how remote it was. The population was just over one thousand. Sketchy phone signal. No cafes, just a chip shop. The nearest town was a two-hour drive away—if you could even call it a town. It was a world apart from anything I had ever experienced.

I was there as part of a corporate program that sent "highly skilled professionals" to contribute their expertise toward Aboriginal economic development.

I arrived full of hesitation. *Who am I to offer advice on something as complex as economic sustainability? What if I say the wrong thing?* I assumed I'd feel even more out of place than I did in the corporate world, where I was constantly questioning whether I was enough.

But within days, I realized that all of those worries were completely irrelevant. The elders weren't interested in polished strategies or PowerPoint presentations. What mattered was understanding *them*—their lives, history, and daily struggles.

I spent hours sitting with people in the community, listening to stories of their resilience, connection to the land, and traditions passed down with quiet reverence. The warmth of their hospitality was humbling. They shared meals, laughter, and insights about a way of life so different from mine.

Somewhere in those early days I stopped overanalyzing whether I was contributing or qualified to be there. I just helped.

The real need wasn't the plan we had been sent to draft (although we still needed to deliver on that). It was far more immediate. The phone lines had been down. There were tech issues to troubleshoot. Some needed help navigating bureaucratic government processes. So, that's what we did. These things weren't in the "scope" of my assignment, but they made a real impact.

And the feeling that came with it was like nothing else.

It's called the *helper's high*[3]—the deep, physiological reward that comes from contributing to something larger than yourself. For the first time in a long time, I wasn't focused on *me*. I wasn't calculating how I measured up. I wasn't obsessed with thoughts of *Am I doing it right?* I was simply being of service.

When I returned to the office after the program ended, something had shifted. I stopped micromanaging how I came across in every meeting and started asking, "Where can I add value here?" I worried less about whether I sounded smart and more about whether I was making a difference. And as it turns out, that's exactly what made me more confident—and more effective.

Getting Beyond Yourself to Accept Yourself

Here's the part many people miss:

Most will spend their entire lives trying to earn their worth. They will chase promotions, perfect their image, and accumulate achievements, hoping that finally, they'll feel like they're "enough."

But self-worth isn't earned. It's already there.

The tragedy is that most people never stop running long enough to see it.

So stop.

Stop obsessing over how you're perceived.

Stop micromanaging every interaction.

Stop overanalyzing every flaw.

You will never think your way into worthiness. Self-acceptance does not live in other people's opinions. It does not sit on the other side of success.

When cultivating and strengthening the Attribute of Acceptance, the practice of self-forgetting changes your orientation so you can let go of your hyper-focus on *Am I enough?* It can help you untether from the rut of doubt and instead operate from a mindset of contribution. Your self-worth is no longer dependent on others' approval or your own insecurities.

Here's what self-forgetting might look like in action:

- You're in a meeting. You start second-guessing yourself, calculating how to sound smart and wondering if your ideas are even good enough. **Self-forgetting reminds you:** This isn't a performance. Don't focus on proving your worth. Instead, focus on moving the project forward. Share the idea. Or better yet, ask the quiet person across the table what they think.
- You're about to post something on social media. You hesitate, editing and re-editing, trying to make it perfect. **Self-forgetting reminds you:** You're not posting for the likes or validation, but to offer something useful and of value. Let it help or inspire someone.
- You're walking into a networking event and everyone seems more impressive. You're mentally scanning yourself: Do I look okay? Will people find me interesting? **Self-forgetting reminds you:** You're not auditioning for anything. You're connecting. Be curious, ask questions, and seek to understand what really matters to the person in front of you.

When you stop over-focusing on yourself, your attention flows outward. When you step out of proving and into serving, the fear fades. You start to recognize that you are deserving of self-expression, love, respect, and success—exactly as you are.

Joseph Campbell eloquently stated, "When we quit thinking primarily about ourselves and our own self-preservation, we undergo a truly heroic transformation of consciousness."[4]

By forgetting yourself, you remember your true value.

And when you remember your true value, you discover the profound gift of self-acceptance.

This realization—that you are *enough*—doesn't come from

achievement, approval, or finally getting everything right. It comes from stepping outside yourself, from shifting your focus away from endless self-evaluation and toward what you can give.

Because the irony is, the more you contribute, the more certain you become of your worth. Not because you've proven anything, but because you've lived it. You've felt the undeniable truth that your presence, your effort, and your impact matter.

TAKE A MOMENT:

*Imagine what you could do
if you fully embraced Acceptance
and practiced self-forgetting today,
this week, and beyond.*

The Second Attribute

AGENCY

CHAPTER 9

Can I Handle This?

The Core Question of Agency

One morning while cutting through the office courtyard at my new banking job, I stopped in my tracks.

Elaine.

She was a school friend I hadn't seen in more than a decade. Back when we were teenagers, she was already a master public speaker. Generally a quiet high achiever, when she stood in front of an auditorium of students her deep conviction could make you believe in *anything* she said—she was that good. I was in awe of the way she radiated confidence and wasn't afraid to use her voice.

A few days after our surprise encounter, we caught up over lunch. But something was off. She seemed . . . well . . . dimmed. As we talked, her worries spilled out, and suddenly, like old times, I was her confidante again.

"Honestly," she said, stirring her coffee like it had wronged her, "I used to feel so confident. But being a lawyer in such a brutal environment—it's changed me. I second-guess myself all the time and rarely speak in client meetings. I never feel like I know enough or that my skills are where they need to be. It's weird how a place you thought was a dream can leave you doubting everything you thought you knew about yourself."

I couldn't believe that *Elaine* was now doubting her right to even be in the room.

When I asked if she'd thought about switching firms, her shoulders sagged. "I don't think I have anything to offer another firm," she said. Four years at a top-tier multinational law firm had somehow stripped Elaine of her belief in her competence, to the point that she doubted *all* of her skills.

One phrase she kept repeating stuck with me: "Do I really have what it takes to succeed here?"

I didn't know it then, but her question would echo in my mind for years.

Even after our conversation ended, I couldn't let it go. Elaine had always been brilliant—objectively successful, clearly capable—and yet here she was, shrinking under the weight of her own doubt. Her story stayed with me because I knew it wasn't rare. I started seeing versions of Elaine in people all around me—especially in high performers who looked confident on the outside but quietly questioned their value.

Though I'd interviewed hundreds of people for research and consulting work, I'd never focused on a cohort of people who, like Elaine, were seen as highly capable by others (and objectively were high performers) but still grappled with deep internal doubts about their abilities. Years later, still turning over Elaine's words in my mind, and now on a mission to understand how self-doubt operates, I reached out to my online audience.

I wanted to see just how widespread this quiet questioning really was. I asked, "Do you ever doubt your capability, even when others see you as competent?"

The response was overwhelming:

"I always hesitate to go for new roles or projects. I keep thinking, 'What if I'm not capable enough to handle it?'"

"Even when there's something I know how to do, I find myself constantly double-checking with others because I'm worried I've missed something."

"I work excessively long hours, not because I need to, but because I doubt the quality of my work and feel I need to keep improving it."

"I talked myself out of taking a promotion I was offered because I didn't think I deserved it. This is despite my leading many high-profile projects and receiving nothing but positive feedback."

"Every time I get something right at work, I can't help but think, *But look at what Jemima's doing, she's way ahead of me.* It's like no matter what I do, I'm always a step behind someone else."

It was crystal clear. Elaine wasn't alone. And if you've ever caught yourself thinking that you don't have what it takes, that you can't handle things, that you're not up to the task, you are like Elaine and so many others.

Why Readiness Matters

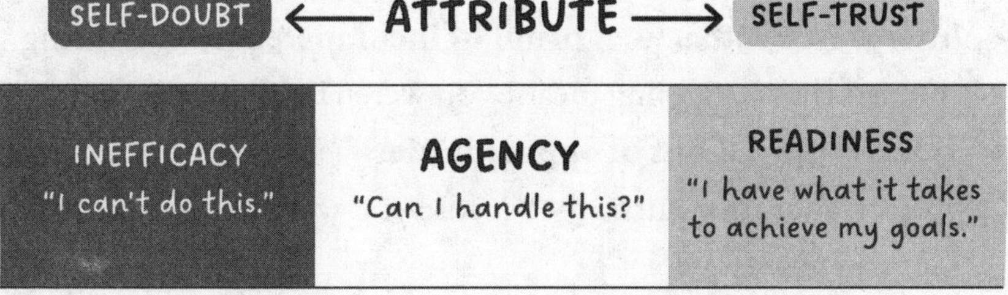

The search for readiness is the defining challenge of Agency, the second Attribute in our Doubt Profiles. When you experience a sense of Agency, you believe that you can figure things out. You not only believe that you have the skills you need, but also *trust* that you have the ability to adapt, learn, and solve problems as they come. When Agency is a vulnerability, self-doubt looms as a nagging sense that you're incapable or ineffective at handling challenges or making progress. You don't just question your skills and abilities—you question your capacity to act at all. Every decision feels like a risk you're not equipped to take.

I've seen highly capable and talented professionals hold back from jobs they were more than qualified for. I've seen people bomb interviews, not due to incompetence, but because they had already decided they weren't going to succeed. Their internal feeling of inefficacy shadowed everything they did. Each rejection or missed opportunity fed the cycle: *See? I knew I couldn't do it.*

Agency-related insecurity seeps into every area of life, telling you you're too undisciplined for that fitness plan, too talentless to try something new, too unqualified to take the next step, too lacking in relationship skills to make a relationship work. So you stay stuck, waiting to feel ready. But that moment never comes.

Think about how many times you:

- **Downplayed your achievements,** chalking them up to luck, timing, or the help of others instead of your own ability.
- **Interpreted setbacks as proof of incompetence,** convincing yourself that struggling meant you weren't cut out for it.
- **Talked yourself out of opportunities**—a promotion, a new project, a creative pursuit—because you weren't 100 percent sure you felt ready.

- **Measured yourself against others' success,** believing that if they were ahead, it must mean you're behind.

When you don't believe in your own Agency, you hesitate. You second-guess. You talk yourself out of opportunities before they even begin. With every "maybe later," you reinforce your self-doubt and continue to stay stuck in a belief that *I can't do this.*

The Templates in Our Minds

Our internal stories about our abilities and how effective we are in the world begin early. As kids, life was one big experiment—first steps, climbing things we probably shouldn't, testing limits, learning by doing. Whether we were supported or discouraged, pushed or held back, our sense of Agency was shaped by the people around us—our parents, teachers, and mentors. Worry and criticism planted seeds of self-doubt. Encouragement and praise for effort planted seeds of self-belief.

Then came adolescence, when your sense of Agency was tested in new ways. You were up against tougher challenges—peer pressure, school stress, and that chaotic cocktail of changing hormones and identity crises. Every win and knockback reinforced a story about what you could or couldn't handle.

By adulthood, those stories tend to harden. It's no longer about climbing jungle gyms or passing a calculus test. Now it's about navigating work, relationships, responsibility, and life's uncertainties. And often, the story of what you're capable of has been playing in the background for so long, you don't even notice it anymore.

In the late 1970s, social cognitive psychologist Albert Bandura conducted a series of experiments to understand how people's beliefs affected their actions. His research found that people who *believed* they could perform well were more likely to take on challenges, persist through difficulty, and succeed—not because they had more talent, but because they had more belief.[1]

So what happens if that belief—your sense of Agency—is shaky? Does that mean you're destined to feel continually incapable, less than, or unqualified? Why do those early impressions stick so tightly? And why is it so hard to imagine that your story could be different?

Imagine your mind as a giant filing cabinet. Inside each drawer are the memories, patterns, and lessons you've collected over time—what psychologists call *schemas*. Each time something new happens, your brain rifles through these drawers, looking for a familiar file to

help you make sense of it. Over time, these stories become your default templates. They quietly shape how you interpret events, people, and, most powerfully, yourself.[2]

Every single experience you've had since birth—whether amazing, humiliating, or somewhere in between—has been filed away in that mental cabinet. If you were encouraged to problem-solve and take risks, your mind is filled with evidence that you're competent and capable. When you face a challenge, that schema kicks in: *I've handled tough situations before—I can do it again.*

Research published in 2021 shows that people with positive self-schemas are more motivated, better problem solvers, and less likely to avoid difficult tasks.[3] In other words, those confidence-boosting files serve as a personalized road map, guiding you through obstacles using the templates shaped in childhood and honed in your teens and early adulthood.

But let's say your childhood was filled with more criticism than confidence-boosting. Maybe you had helicopter parents who panicked excessively when you faced challenges, or teachers who were quicker to point out what you got wrong than what you got right. That filing cabinet might be packed with memos that tell a different story. Instead of seeing obstacles as things to overcome, you see them as threats. As an adult, when a challenge arises, those negative schemas don't sit there quietly gathering dust. They pop open at the worst possible moments—when you're facing a new challenge, applying for a job, or speaking in a meeting. They start screaming, *You can't handle this! Retreat!*[4]

While your sense of Agency has become deeply ingrained, your filing cabinet isn't welded shut. You can edit what's inside the drawers. You can reorganize, rewrite, and replace the files.

The Role of an External Perspective

Perhaps ironically, highly capable people are often the worst at naming and taking ownership of their strengths. When I ask leaders we work with, "What are you really good at?" I'm often met with blank stares. They fumble around, listing generic skills, unsure of how to articulate what makes them stand out. They don't see their abilities as special. They're too close to them.

Counterintuitively, a little self-doubt related to your Agency is often a sign of capability. In fact, research has shown that people with lower abilities often think they perform *far better* than they objectively do. It's called the Dunning-Kruger effect. You might have heard of it.

For example, people who score lower on emotional intelligence often overestimate their performance on emotional intelligence tests. Likewise, many people who are lacking in financial literacy skills have been shown to overestimate how good they are at managing their money.[5] Meanwhile, those who are genuinely competent often underestimate themselves. Why? Because as David Dunning, one of the researchers behind the effect, puts it, "Those who lack expertise lack the expertise to know they lack the expertise."[6] Basically, the more skilled you are, the more aware you become of what you *don't* know.

That's why it's so important to have people around you who can reflect your strengths back to you. When the people around you see your potential, and encourage that potential, it actively builds your sense of Agency. Researchers call this *verbal persuasion*—the impact of hearing others affirm your strengths.[7] If you're surrounded by people who doubt you or are quick to criticize, it's easy to let that doubt seep into your mindset and start doubting yourself. This is the Golem effect: Low expectations lead to low performance. It limits your growth and effort.

A toxic workplace, hypercritical boss, micromanaging client, or team that undervalues your contributions can erode your confidence over time and reshape how you see yourself. Psychology researchers call this *career imprinting*[8]—where professional experiences shape your long-term belief in your abilities. That's exactly what happened to Elaine.

Four years in a cutthroat law firm had rewired her mental filing system and made a mess of her folders. The same woman who once commanded stages now struggled to speak in public, convinced she had nothing to offer.

That's one reason why work culture matters so much. The environment you spend most of your waking hours in either builds you up or slowly breaks you down. A toxic culture can rewire your brain, leaving you with files you never chose that you nonetheless now find yourself constantly referencing.

But the opposite is also true.

When you're surrounded by people who see your capabilities and potential—and actually tell you so—something incredible happens. Their belief lifts you up, inspiring you to aim higher and stretch beyond what you thought was possible.

Research has shown time and again that when leaders set high expectations (and reinforce them with encouragement, support, and praise), teams tend to rise to the occasion. They aim higher, work harder, and achieve more.[9] The same phenomenon happens in classrooms. When teachers genuinely believe in their students, those students often excel and get better grades.[10] This is the power of the Pygmalion effect. When others see greatness in us, they reflect back what we might not yet see in ourselves. They remind us of our capabilities and give us the confidence to believe in them too.

The problem is, many of us don't have someone doing that for us. And that's where your own mind can help . . . or hurt.

Your mind works like the YouTube algorithm. It curates content based on your viewing history—what you already pay attention to—thanks to your brain's Gatekeeper (the internal filter that decides what gets your focus and what doesn't, based on past patterns you've learned). If your internal playlist is a highlight reel of your failures, that's what plays on repeat. But you *can* change the channel.

Start reviewing your mental files. Look for the ones that no longer serve you. Add new ones that reinforce your abilities. When moments arise that shake your confidence or make you question whether you really have what it takes, these new files become your counterweight—reminders of what you're capable of.

I like to think of it as having a mental (and actual) highlight reel. Think back to times you succeeded, big or small—moments of courage, problem-solving, or resilience. Everyone has succeeded at *something* at *some point* in their life. Write those moments down in a journal or a folder on your desktop. With clients (and myself), I take it a step further: Gather real evidence. It might be the thank-you email from your top client, the "Great job" handwritten card from your boss, and the teammate saying, "You handled that really well." Save it all. Sometimes others see strengths in you that you've stopped noticing.[11] You can even ask a trusted friend or mentor what they think you do best and add their notes to your collection.

This becomes your go-to playlist whenever your Agency needs a lift. The more you replay it and remind yourself of what you *have* accomplished, the more confident you become in what you *can* accomplish. You're intentionally building a new mental filing cabinet stacked with proof of your capability, ready to be pulled open the next time doubt tries to take over.

It's exactly what happened with Elaine. When we reconnected, I reminded her of the powerhouse she used to be—how confidence had once come so naturally to her. And slowly, she started reshuffling her mental files and building that playlist. Six weeks later, she emailed me:

Your words really struck a chord. I've been reminding myself of what I do well, and it's changing how I feel. For the first time in years, I feel a little like my old self again.

The Promise of Agency

Life is full of uncertainty. When your sense of Agency is strong, you don't expect or need to have all the answers up front. You simply trust that you can figure it out. You might not have all the tools just yet, but you believe that you can learn, adapt, or find someone who can help. It's the self-belief that comes from having a can-do attitude, backed by a commitment to taking action. Where the Attribute of Acceptance is believing you're *worthy* of success, Agency is believing you have the *ability* to make success happen—or whatever your goals in any moment might be. When Agency is strong, you're driven, disciplined, and ready to get your hands dirty. And because you expect yourself to succeed in the long run, you take more action—making success more likely.[12]

In the context of Agency, your actual abilities matter, but your *belief* in them matters more. Why? Because belief drives action, and action creates results. No belief in your abilities leads to no action. No action leads to no progress.

Agency isn't blind optimism. It's not: *I'll succeed the first time.* A belief in instant success is not confidence; that's a form of delusion. Agency is the ability to pivot and grow from setbacks. When things go sideways (and they will), you don't freeze or quit. You regroup, adapt, and find new ways around the obstacles—or even create an entirely new game plan altogether.

PRACTICE: Assess How Agency Is Showing Up for You

When you understand the role Agency plays in your Doubt Profile, you get a clearer sense of how much you trust your own abilities—and where that trust might need rebuilding. It gives you the insight you need to start taking action, one habit at a time, to strengthen your belief in what you're capable of.

Look back at your Agency score from the Doubt Profile diagnostic. What was it?

Agency Score: __________ Zone: ________________________

(For Zone, write in "Red Alert," "Hindrance," "So-So," "Hidden Strength," or "Superpower.")

Use these prompts to dig deeper. You might like to journal your responses:

What are your vulnerabilities around the search for Agency?

- **Think back to your early years—what messages did you receive about your abilities?** Were you encouraged to try, fail, and learn? Or were you taught to play it safe and avoid mistakes?
- **How does inefficacy show up in your life?** Do you hesitate and second-guess yourself when faced with challenges? Avoid opportunities because you're unsure you have what it takes?
- **When was the last time you took action, even when you didn't feel fully ready?** What happened? What did you learn about yourself?

- **How might your life feel and look different if you fully trusted in your ability to figure things out, even when you don't have all the answers yet?** What doors might that open for you?

When Agency feels impossible, what strength can you lean on?

- If you struggle to take action because you feel you're not "ready," lean on **Acceptance**. Give yourself permission to act before you feel perfect or fully prepared. (Refer back to the Acceptance Attribute for insights and reminders.)
- If fear of external outcomes is paralyzing you, tap into **Autonomy**. What's one action you can take to show yourself that you have a role in how you respond and can help shape what happens? (You'll learn more about Autonomy when you get to the third Attribute.)
- If fear is keeping you frozen, anchor yourself in **Adaptability**. What's one immediate action—a breath, a break, a reframe—you can take to regulate your emotions and get unstuck? (You'll learn more about Adaptability when you get to the fourth Attribute.)

• • • •

In chapters 10 through 12, we'll explore the biggest threats to Agency and ways to counter them: imposter beliefs that whisper you're a fraud, the comparison trap that makes you feel like you're falling behind, and the instinct to avoid taking action when you don't feel confident. These patterns are reinforced by mental habits that chip away at your self-belief. But those habits can be changed.

You'll learn practical tools and mindset shifts to help you build new patterns of thought and behavior—ones that strengthen your sense of Agency and help you take bold, meaningful action. Then, in chapter 13, we'll dive into a mind-expanding idea to help you rebuild and fortify your sense of efficacy for the long haul.

CHAPTER 10

Am I an Imposter?

Explanations (and Excuses) Are Thoughts, Not Truths

When Pauline Clance was a child, she was always second-guessing herself. After nearly every test she took (and aced, mind you), she'd tell her mom, "I think I failed." She was the first in her family to go to college, eventually earning a PhD in psychology. But even then, her nagging self-doubt taunted her, telling her that she had only tricked people into thinking she was smart.

When Pauline met Suzanne Imes, a colleague at Oberlin College, the two bonded over their shared talent for doubting themselves, despite being highly accomplished. Intrigued, they decided to dig deeper. Over five years, they interviewed more than 150 successful women (who had earned PhDs, were respected professionals, or were academically outstanding students). Most of them confessed to feeling like "frauds."[1] The women were, in fact, part of the first generation of women taking their place in the modern workplace (or preparing to).

What Clance and Imes discovered, and published in a groundbreaking 1978 paper,[2] was as fascinating as it was depressing. Many of the women believed the reason for their success was due to being in the right place at the right time, knowing the right people, or simply their ability to charm their way to the top. Did they attribute their success to hard work, intelligence, and skill? Nope, just a cosmic fluke.

If you've ever felt like a fraud, like you don't truly deserve your success, congratulations—you're human. Studies show that up to 82 percent of people have experienced these feelings at some point.[3] Even Einstein couldn't escape it. He once said, "The exaggerated esteem in which my lifework is held makes me very ill at ease. I feel compelled to think of myself as an involuntary swindler."[4]

Yes, *that* Einstein. The man who fundamentally changed the way we understand the universe. If the genius mind who figured out relativity felt like he was faking it, it's no wonder the rest of us feel that way at times. When your sense of Agency is shaky, the stakes are high, or things aren't going as planned, it's easy to slide into the mental trap of *I don't have what it takes.* Never mind all the evidence that says otherwise.

Clance and Imes called this the "imposter phenomenon," and described it as a perceived sense of intellectual and professional fraudulence and an internalized feeling of being undeserving of one's achievements, despite a strong track record of success. And it applies to men too.[5]

When You Feel Like an Imposter

Over time, the imposter phenomenon became popularly known as the "imposter syndrome." Words are powerful, and this rebranding changed how we perceive the concept, essentially pathologizing it into something that sounds like a medical condition. People often "self-diagnose" imposter syndrome, because it can be comforting to have a label for this aspect of our self-doubt. But hiding behind it? That's when it gets in your way.

At its core, the concept of the imposter phenomenon reflects a distorted self-perception and belief that you don't have the genuine skills, intelligence, or qualifications that others believe you to have. Organizational psychologist and professor Adam Grant sums it up beautifully with this powerful paradox: "If you doubt yourself, shouldn't you also doubt your judgment of yourself? When multiple people believe in you, it might be time to believe them."[6]

Signs of feeling like an imposter include downplaying your achievements (*It's no big deal—anyone could have done it*) and magnifying your flaws (*Sure, I got the promotion, but I suck at public speaking, so it doesn't count*).[7] This distorts how you see yourself, and can lead you to excessive effort to *prove* yourself. You try to overcompensate (*If I work twice as hard as everyone else, no one will notice I don't belong*). The irony, though, is that this relentless drive to overwork[8] and chronically overperform[9] doesn't even tip the scales in your favor. In a meta-analysis looking at ninety-seven different samples of people, researchers uncovered an eye-opening truth: Those who overwork might clock more hours than their peers, but they don't perform any better.[10]

The inner imposter can also lead you to avoid the risk of "exposure." Your inner voice whispers, *Don't even try; you'll just fail spectacularly and everyone will find out you're a fraud.* So, you keep your mouth shut in meetings or shy away from new challenges. You might avoid taking pivotal career moves, because if you don't try, there's no risk of failure. You might overreact to feedback (seeing it as an attempt to "expose you") or hesitate to ask for help because you don't want to appear incompetent or inadequate.

The actor Jason Segel may be best known for his role in *How I Met Your Mother,* but in the lead-up to his directorial debut with the show *Dispatches from Elsewhere,* he was overwhelmed with the feeling that he was there by mistake. Instead of hiding his anxiety, he owned it and said, "This is my first time directing—if I do anything that bugs you, let me know. It is not intentional; I'm doing this for the first time." He later reflected, "That was freeing. You say it out loud, and it breaks the ice."[11]

PRACTICE: Reframe Your Imposter Thoughts

Imposter thoughts feel real, but they're often just distorted beliefs. Reframing your thoughts—without trying to eliminate them—is proven to help you replace old narratives with more constructive ones that support action, learning, and self-trust.[12]

Step 1: Catch the Thought

Notice the imposter narrative running in your mind: "I don't belong here." "They're going to realize I'm not as smart/capable as they think." "Everyone else knows what they're doing except me."

Step 2: Reframe It

Now, rewrite the story with a lens of growth, curiosity, and truth. Here are some examples of powerful reframes:

- "I don't belong here." → "This is an opportunity to grow into the space I've earned."

- "I have no idea what I'm doing." → "I'm in the process of figuring this out—just like everyone else once was."
- "I'm not as smart as they think." → "They see potential in me, and I'm learning to see it too."
- "I got lucky." → "Luck might have opened a door, but I walked through it and showed up."

Step 3: Say It Out Loud (or Write It Down)

Reframing works best when you give the new thought airtime. Speak it. Write it. Let your brain hear the truth from you. Repetition rewires.

Bonus Tip: Use "Yet" to Shift into a Growth Mindset

Whenever you feel like you're falling short, try adding the word *yet* to the end of your sentence: "I'm not confident speaking up . . . yet." "I don't understand this concept . . . yet." "I haven't figured it out . . . yet." This deceptively simple strategy comes from renowned psychologist Carol Dweck,[13] whose research on growth mindset shows that believing you can improve with effort is one of the most powerful drivers of success. "Yet" signals that growth is still possible. It keeps the door open and reminds your brain that your current struggle isn't your final destination.

Remember, imposter thoughts are fueled by beliefs, but beliefs can be rewritten. And reframing is how you begin.

Elevating Hidden Expertise

We get caught up in feeling like an imposter when all our focus is on how we're viewed by others. We zoom in on what we *can't* do, or we judge our abilities negatively—at least more negatively than the people around us, who we are convinced believe we're more intelligent or capable than we truly are. We believe that others are more competent, faster, more efficient, or that they "get it" in a way that we don't. Not only is this usually not true, it also overlooks all the unique capabilities we bring to the table.

In a study published in 2022, a researcher at MIT found that in their quest to overcompensate, people with imposter thoughts actually tend to be rated as having better interpersonal and social skills, like empathy, listening, and being able to relate to others.[14] Basically, you might feel like a fraud, but others think you're fantastic to work with. This was definitely true for me, and something I didn't even realize at the time.

When I was in my first few years in the banking industry, especially as I started to be offered more senior roles, I absolutely felt like an imposter, that I was one question away from being exposed. But I also knew that I was *very* good at relationships. I was a confident communicator, great at navigating conflict, cooperating, encouraging others, relating to people at all levels—and these skills were highly valued. In a way, I tried to overcompensate on an interpersonal level. The upside was that these very skills that came naturally to me also made me stand out for the right reasons—even though I was spiraling internally. And that upside, that strength, is something I now often call on to rise above my self-doubts.

I love to tell a story about the legendary American graphic designer Paula Scher when she was hired in 1998 to redesign the logo for Citibank after its merger with Travelers Insurance Company. Scher has explained, "After we had the first meeting, I drew the Citibank logo on a napkin and walked out." As she got up to leave the room, someone from the Citi team asked, "How can it be that it's done in just a matter of seconds?" Without missing a beat, Scher confidently replied, "It's done in a second and thirty-four years." She continued, "It's done in a second and every experience and everything that's in my head."[15]

That client's surprise aligns with how society often undervalues expertise gained over time, mistaking years of honed skill for mere minutes of effort. And this is something we do too—we undervalue the experience, skills, qualities, late nights, lessons learned, and Attributes we've developed over time. We undermine our own Agency.

The lesson here is that, like Scher's logo design (that she was paid $1.5 million for), your competencies go far deeper than what's visible at first glance. And yet, we often downplay or dismiss them. If something comes easily to us, we often assume it must not be valuable ("Oh, this? It's nothing."). We fail to recognize that all the years of experience, learning, and effort are what make us skilled, not just the outcome in the moment. That's the catch with the imposter phenomenon—it tells us that unless we're constantly pushing ourselves to the limit or struggling, our skills don't count. Or worse, that we've somehow "gotten lucky" instead of earning our place.

Not only that, the skills you develop in one domain can be transferred to others in ways you might not have expected. This is how a corporate architect became one of Nike's top shoe designers. Back in 1985, Nike needed fresh creative talent, so they held a twenty-four-hour shoe design competition. Tinker Hatfield, with exactly zero experience in footwear design, entered—and won.

For his first official shoe design, Hatfield drew inspiration from a building he'd studied in architecture school: the Centre Pompidou in Paris. This is an "inside-out" building, where its structural and mechanical systems are all exposed. Hatfield thought, *Why not do that with a shoe? Let's cut a hole in the side and show what's in the shoe.* And just like that, the Air Max 1 was born. It ended up being a huge success, becoming one of Nike's most iconic sneakers. Hatfield says, "To this day, Phil Knight says I saved Nike." He continues, "When you sit down to create something . . . what you create is a culmination of everything you've seen and done previous to that point."[16]

The lesson here is that even when you don't have direct experience in a new role or industry, the technical skills you've developed in another area can still play a crucial role. Hatfield didn't let the fact that he wasn't a trained shoe designer stop him—he used his architectural expertise to bring a fresh perspective to footwear.

If you're feeling like an imposter because you don't have every specific technical skill for your current role, remember that the skills you've developed elsewhere (in other roles, industries, or even outside of work) can transfer in powerful ways. Don't underestimate the value of your perspective or the depth of your experience, even if it doesn't look like a perfect match on paper.

Success Doesn't Demand Expertise

Whether you're a seasoned pro or stepping into something completely new, you're going to face situations where you don't yet have all the technical know-how. A new job, a complex project, or even trying to assemble IKEA furniture without cursing the Swedes—these moments can feel daunting. In our own work and with our clients, Fayçal and I lean on what we call a philosophy of "Essence over Expertise." The idea is simple: Your core "essence qualities"—like resilience, adaptability, and curiosity—often matter more than the technical skills you have in the moment. These are the qualities that help you learn, problem-solve, and navigate uncertainty.

Of course, your expertise matters. So does being competent. These can be built. But the "essence" of who you are at your core—your character and how you show up—often matters more when it comes to achieving success.[17]

These essence qualities are some of the most overlooked strengths we have. Back when I did a short stint in recruitment during my university days, I conducted thousands of telephone screening interviews, mostly for train driver positions. Of the tens of thousands of applicants, *nobody* had ever been a train driver before. But we weren't looking for prior experience; we were looking for what they called the "potential to learn." Translation: Was the person adaptable and resilient? Could they figure things out without derailing an actual train? Those essence qualities were the real drivers of success—pun totally intended.

It's the same with any new challenge you take on. Whether you're launching something new, switching industries, or stretching into a bigger role, your creativity, resourcefulness, and determination will often carry you further than technical expertise alone.

When you focus on what you *know* that you have—your character, your grit, your capacity to grow—you build the kind of self-trust that doesn't rely on having all the answers before you begin. You strengthen your Agency: that inner belief that you can figure it out, *even* in areas where your current expertise or competence is still developing. You already have what it takes, even if it's still taking shape.

PRACTICE: Use "Essence Qualities" to Build Expertise

We all have the potential to shift our relationship with self-doubt and elevate our sense of Agency by tapping into our essence qualities. Try this three-step exercise that I use to support clients who are grappling with imposter thoughts.

Step 1: Reflect on Achievements and Identify Essence Qualities

Take a moment to bask in your glory (yes, *your* glory). Think back on your achievements—personal or professional, big or small. Now, zero in on the *essence qualities* that got you there. Forget about the shiny outcomes you probably think you didn't deserve anyway. Focus on the *process*; the effort and qualities you applied to get there.

Here are some examples:

- You successfully led a team to complete a major project on time and within budget? Your essence qualities might be leadership, perseverance, and problem-solving.
- You mastered a new software tool that improved your productivity? Your essence qualities could be curiosity, flexibility, and persistence.

- You successfully organized a community volunteering event at an animal shelter? Your essence qualities may be creativity, collaboration, service, and initiative.

Remind yourself that these are part of your "personal essence toolkit." They don't just disappear because you're doubting yourself. They're *always* there, waiting for you to call on them.

Step 2: Detect Your Gaps

Next, identify the "gaps"—these are the skills or competencies you feel you lack or need to develop, and they're often the things that contribute to your feelings of being an imposter. Reflect on times you felt out of your depth. Was it during a meeting where someone rattled off industry jargon like they were born in a PowerPoint slide? Or maybe you felt inadequate tackling a task you weren't fully trained for? What specific skills or knowledge did you feel you were lacking in those situations? Write those gaps down.

Step 3: Bridge Gaps by Applying Essence Qualities

This is the fun part. Here, you pair your essence qualities from step 1 with your identified gaps from step 2 in a way that reminds you that you can leverage your existing qualities to fill these gaps. This not only builds your confidence but also creates a practical plan for growth. For each gap, think about which of your essence qualities from step 1 can help you address it. Then, write down specific actions you can take.

For example:

- "To improve my data analysis skills, I'll use my persistence and curiosity to explore new data tools and take extra time to practice."

- "To address my gaps in industry knowledge, I'll harness my love of learning to read industry reports and my curiosity to attend industry conferences and webinars."

Now you have a personalized plan for growth. You're acknowledging the wealth of essence qualities that have already helped you get to where you are and will continue to carry you forward. Think of this as your *Achievement Portfolio*—not unlike the highlight reel you began building earlier. It's a growing collection of evidence that reinforces your Agency: your capacity to learn, adapt, and rise. Beyond boosting confidence, it's also a practical reference for interviews, promotion conversations, or moments when imposter syndrome creeps in. Most importantly, it reminds you that your value isn't fixed to a title or role. It lives in the transferable strengths and qualities you bring with you, wherever you go.

• • • •

Success isn't reserved for those who have it all figured out. It belongs to those who keep showing up, even when they doubt their abilities. Imposter thoughts will try to convince you that you're only here because of luck or timing—but don't forget the truth: You're capable. You've worked for this. You've grown for this. You're not a fluke. You're the product of persistence, courage, and a willingness to learn. You don't need to know everything right now. You just need to trust that you can figure it out.

CHAPTER 11

Why Is Everyone Better Than Me?

The Curse of Comparison-itis

Nathan was exhausted. Even through the screen, I could see the tired eyes of someone who felt like he was constantly falling behind. Just five weeks into his new role as a project manager at a fast-moving tech company, he was already teetering on burnout.

He had landed the job after a grueling five-round interview process, backed by a glowing recommendation from a former colleague. But despite that, his feelings of self-doubt came pouring out in our very first virtual session together.

Nathan came to me for support in making the transition, but it was clear he felt lost in his new role when he explained, "Everyone else seems to have it all figured out. Why can't I pick things up more quickly?"

As we talked, I noticed how much weight he was giving to every single comment from his colleagues and managers. He was taking everything to heart, internalizing it, and letting it define him. When I tried to highlight the positives he'd mentioned, he dismissed them outright, as if they didn't count or weren't enough to matter.

Then came Nathan's gut punch. An old friend of his from high school had moved to New York for a new opportunity, and Nathan saw on LinkedIn that he'd landed a huge client deal and gotten promoted. "It's only been a few months," Nathan said, "and I feel like I'm already falling behind."

Comparison had a chokehold on him. Not just in work, but in life. When he was out with friends, he found himself tallying up their accomplishments—promotions, investments, big trips—and measuring them against his own. At first, he felt happy for them. But then, almost automatically, the doubts would creep in. *Why am I not traveling as much as they are? Should I be making more money? Investing in real estate like they are? Why don't I have my life as figured out as they do?*

That's the trap of *comparison-itis*—the destructive habit of measuring ourselves against someone else's highlight reel.[1] Maybe it's a friend's photo of themselves lounging on the French Riviera, or news that your cousin just bought a four-bedroom house with a pool while you're still Googling "cheap ways to soundproof an apartment."

The truth is, you cannot measure your way to self-confidence.

"Measuring Up" Is the Wrong Goal

Comparing ourselves to others has a function. From an evolutionary perspective, it helped us figure out where we stood in the group. Are we doing okay? Are we safe? Psychologist Leon Festinger named it *social comparison theory*. The idea is that in the absence of objective markers, we instinctively use other people as mirrors to evaluate our own abilities or progress. This habit of comparing yourself to others is often an instinctual part of your daily life.

The problem is, when you're already struggling with doubt and questioning whether you have what it takes to handle your career and goals, your brain isn't looking for fair comparisons. It's looking for confirmation. It looks for evidence to support your beliefs. And what better proof than finding all the ways you're falling short compared to others?

You don't just measure your achievements on their own merits—you measure them against the achievements of *everyone* around you. Your high school friend whose start-up just went public. The fitness guru who somehow has a six-pack and a full-time job. This is only amplified at work. You're under constant scrutiny. You're subjected to performance reviews, KPIs, peer feedback, and screen monitoring with activity trackers, all designed to evaluate you, over and over again. It can turn work into a competition of "high performers" versus everyone else. If you struggle with agency, this constant comparison does one thing: It turns the spotlight on everything you lack.[2]

Two studies, one from 1995,[3] and a replication in 2021,[4] found that Olympic bronze medalists in individual competitions like swimming are often *happier* than silver medalists. How can that be? How can someone who's trained so intensely feel better about being third than second?

Comedian Jerry Seinfeld explained it most memorably in one of his stand-up routines: "Think about it," he said, "if you win gold, you feel good. If you win bronze, you think, 'Well, at least I got something.' But if you win silver, it's like, 'Congratulations, you *almost* won. Of all the losers, you came in first in that group. You're the number one loser. No one lost ahead of you.'"[5]

This tendency of silver medalists (the "almost winners") to compare upward is similar to what happens for all of us when we focus on what *could have been* or what we missed out on. The "if only" mindset feeds regret, disappointment, and self-criticism.[6] It's called *upward counterfactual thinking.*

There's a glaring problem with always trying to "measure up"—trying to be the "best" or the "gold medalist." It sets you up for a life of stress and exhaustion.

Why? Because no matter how much you achieve, there's always someone else out there who has done more. And you have to keep pushing—keep achieving—just to hold on to any sense of success. If you *can* keep achieving, then you are "successful," a "champion," a "winner." But the moment you stop? You're no longer successful. You're a loser. A failure.

This is what Nathan was contending with, and it's why his self-doubt was so relentless. It's where the chatter started to flood in: *I'm not good enough. I'm so far behind. I can't keep up.* And so begins the cycle of anxiety, self-doubt, and endless comparison.

Confidence doesn't come from tallying up wins or making sure you've got more gold stars than the next person. It comes from trusting your own path, even when it looks nothing like anyone else's.

Run Your Own Race

When track athletes compete, they don't waste energy glancing at the competition. They keep their eyes on the finish line, dialed into their own race strategy.

When I shared this racetrack analogy with Nathan, he nodded before I could even finish my thought. It turned out he was a collegiate track athlete, so the metaphor clicked instantly. "In a race," I asked, "how do you stay focused on your lane and not get thrown off by what's happening around you?"

"I'd concentrate on my form, pace, and breathing, and I'd block everything else out," he said.

"Exactly," I replied. "What if you took that same approach with your new job? Focus on your own progress?" Nathan understood this principle from years of running.

The same mindset of running your own race applies to any high-pressure environment, including the water. Olympic swimmer Michael Phelps, one of the greatest athletes of all time, never spent his energy worrying about the guy in the next lane. He didn't psych himself up by imagining he was *better* than his competition—he didn't focus on them at all.[7] Instead, he made a point of preparing for the unexpected.

In his hours of training, he visualized everything that could possibly happen during competition—his goggles snapping, his cap coming loose, or a late start, and he developed a plan for each scenario. Each of these undesirable situations was "programmed in his nervous system" so if it happened, he wouldn't panic or get distracted by what the guy in the next lane was doing. He would just execute his plan.

Phelps understood that comparison was a distraction. Your best chance of success doesn't come from trying to outshine someone else. It comes from being so grounded in your preparation that nothing shakes you.

Nathan wasn't racing in the Olympic pool, but the same principle applied. His confidence didn't need to come from matching his colleagues stride for stride. It needed to come from focusing on his own lane, his own pace, and preparing for what was in front of him.

For Nathan, confronting the stress and competitive atmosphere of his new job made it even harder to stay grounded. But the more he tried to practice this mindset, the more he started noticing how often his brain went into *compare mode*. It was almost like a reflexive muscle twitch of self-doubt. A colleague sharing a big win? *Trigger.*

A manager casually praising someone else's work? *Trigger.* A friend mentioning their new investment? *Trigger.*

But now he had a plan. Instead of spiraling into rumination, Nathan began pulling his focus back—back to *his* own work, *his* own progress, *his* own pace. He wasn't trying to silence the comparisons completely, he just chose not to let it drown him out.

Only a month into his "stay-in-your-lane" approach, Nathan was feeling more grounded, more confident, and—most importantly—less reactive. During a team meeting, as his colleagues shared their updates, he felt that familiar twinge of comparison starting to creep into his gut. This was a usual trigger for him. His thoughts started racing: *Why didn't I do something like that? Am I falling behind?*

But this time, he caught himself. He took a deep breath, refocused on his own work, and stayed in his lane.

When it was his turn to speak, Nathan didn't try to inflate his update or outshine anyone else. He simply shared his progress on a project he'd been managing. And go figure, nobody questioned his competence. Nobody cared about how his work stacked up against anyone else's. By keeping his eyes on his own lane, Nathan was able to contribute confidently without letting his insecurities hijack the moment.

That's the beauty of focusing inward: Progress isn't about matching someone else's stride. It's about taking *your* next step. Or as my mother used to constantly remind me, "The only comparison worth making is with who you were yesterday."

PRACTICE: Gratitude Refocus

Comparison narrows your focus and zeroes you in on what you *don't* have. To pull yourself back into your lane, shift your gaze toward what's already good in your life—there's always something to appreciate. Maybe it's the supportive coworkers who have your back, a job that stretches you, the paycheck that funds your adventures, or the people who love you unconditionally. It could even be as simple as appreciating a warm coffee on a chilly morning or the sound of the wind through the trees. Research has found that gratitude boosts your belief in your ability to achieve your career goals.[8]

So every morning, finish this sentence: "I am grateful for ______________________."

Whatever comes to mind, let that gratitude ground you. Gratitude helps you steer clear of comparison, pulls you back to your own path, and reminds you to embrace what's already yours.

Less Comparison, More Emulation

When you stay in your lane and you're not busy side-eyeing everyone else, you can actually run your own race. And sometimes, if you're intentional, looking at someone else's path can even help you run it better.

Every time you see someone succeed—especially someone similar to you or someone you admire and want to emulate—you're faced with a choice. You can let that success poke holes in your self-belief, highlighting everything you think you lack. Or you can use it as fuel. You can let it motivate you.

If this person you respect has done it, why not you? Research backs this up: Seeing people who we consider similar to us achieve their goals can crank up our motivation and strengthen our sense of agency.[9] It's the reason mastermind groups and mentorship circles are so powerful—they surround you with people who remind you of what's possible.

When Nathan started seeing others' successes as a blueprint instead of a judgment, something shifted within him. He stopped asking, "Why am I so far behind?" and instead started asking, "How did they do that, and what can I learn?" That's the magic of turning comparison into emulation.

Comparing yourself to them can shine a spotlight on all the ways you fall short, undermining your confidence. Instead of letting envy gnaw at you and thinking, *I'll never be as skilled/successful/wealthy/fit/whatever as them,* shift to *Their success shows me what's possible and motivates me to grow in my own way.*

Learning from others is incredibly valuable (and science says it works), but remember this important distinction: You are *not* them. The key is *intentionality*. Your journey comes with its own set of challenges, milestones, and detours. Their path can teach you, sure, but it doesn't define what *you're* capable of. When you approach it this way, you unlock the best of both worlds: learning and inspiration from others, coupled with a laser focus on what matters most—*your* own growth and unique journey.

PRACTICE: Stay in Your Lane—Prepare, Don't Compare

When we constantly compare ourselves to others, it's easy to get distracted by what they're doing—how they're performing, achieving, or winning. But staying in your lane means focusing on *your* journey, not theirs. It's about self-improvement over comparison. And one of the most effective ways to ground yourself in that focus is through preparation. The more prepared you are, the less you'll feel the need to compare. This leads to a stronger sense of Agency, where the anxiety that drives comparison begins to fade.

As writer, producer, and director Aaron Sorkin of *The West Wing* fame says, "Before I can write a single scene, I badly need to know the intentions and obstacles . . . That point of friction—somebody wants something, something stands in their way of getting it—that's the drive shaft of drama. And then the tactics used to overcome that obstacle or series of obstacles standing in their way—that's the clothesline that your character hangs on."[10]

Just like in a story, staying in your lane means knowing what you want (your intention), and anticipating what could go wrong (the obstacles). By doing this, you don't get derailed and distracted by someone else's lane.

So whether you're gearing up for a presentation, an interview, a high-stakes negotiation, or any situation where you want to perform at your best, preparation is your best tool to stay focused. Here's what you do:

Step 1: Anticipate Obstacles and "Unlucky Scenarios"

Ask yourself, *What challenges have others faced on their journey? And what challenges am I likely to face? What could possibly go wrong?* These could include handling disapproval from your family or co-workers, dealing with imposter thoughts, or even coming up against intense boredom and getting stuck in a procrastination spiral. For instance, if you're anxiously worried about a presentation going horribly wrong, try imagining having a mind-blank, a PowerPoint malfunction, or your boss looking unimpressed.

Step 2: Build Your Recovery Plan

Once you've identified the obstacles, figure out how you'd recover. For every "What if?" scenario, come up with a clear "Then I'll . . ." response. These are your contingency plans, or what behavioral researchers call "implementation intentions." For example:

- "If my mind goes blank during the presentation, then I'll pause, take a deep breath, and check my notes."
- "If I mess up the intro, then I'll make a joke about it and keep going."

Write them down. Research shows that having these if-then plans improves performance because you're not scrambling in the moment—you've already got a plan locked and loaded.[11]

Step 3: Visualize

Here's where it gets fun. Mentally walk yourself through the challenges you've identified and practice your response. Picture the moment your boss raises an eyebrow or the Wi-Fi cuts out mid–Zoom call to your investors. Then imagine yourself calmly handling it. This way, when you face a challenge, you can say, "I've been here before, and I know what to do." This mental rehearsal (both the tough moment and the recovery) can help you program it into your nervous system so you feel like you have more control over the situation. You can stay in your lane.

When Michael Phelps's goggles filled up with water in the 2008 Olympics during the 200-meter butterfly finals, he stayed calm. He didn't let the moment shake him because he was prepared. He didn't compare where he was to competitors in other lanes or let doubts creep in. He stayed in his lane, literally and figuratively, trusted his training, and, as he puts it, "swam blind for 175 meters out of a 200 fly, won gold, and broke the world record."[12]

• • • •

Comparison is inevitable, but it's how you use it that matters. You can let it drain you—making every success around you feel like proof that you're behind. Or you can use it as fuel. *If they've done it, why not you?*

The moment you stop measuring yourself against others, you free up energy to focus on what truly matters—your own growth. Because success isn't a race with a single timeline or pace. Everyone has different starting lines and obstacles. The only race that matters is yours.

So take a breath, refocus, and remember: Your job isn't to keep up. It's to keep going.

CHAPTER 12

Do I Just Need to Be More Confident?

"Ready" Is Overrated

Payam Zamani had never even been online when his brother called him in 1994 with an idea that would change his life.

"Honda does not even own a website," his brother said. "What if we built a site where people could research cars and connect with dealerships?"[1]

Zamani had no background in tech. No experience building a business. No road map for how any of this would work. But something about the idea grabbed him. The internet was still uncharted territory—why not them?

With no money for a logo, he painted a graphic designer's garage in exchange for one. His brother built the website in his spare time, while Zamani hit the pavement, pitching dealerships on an opportunity they didn't understand.

"We have no idea what you're talking about," they told him. They didn't think people wanted to look at cars online.

More than 150 rejections later, most people would have quit.

Zamani didn't. Instead, he doubled down. He studied sales, blasted Guns N' Roses to psych himself up, and kept showing up. When his first dealership finally signed on, he knew—if one would say yes, others would follow.

Five years later, in 1999, his company AutoWeb went public at a $1.2 billion valuation. Not bad for someone who had arrived in the US just nine years earlier as a refugee with $75 to his name.

The Lie of "Ready"

Now, suppose I could give you two options about how to live your life. The first is to take action when you feel confident, psyched up, or the timing feels right. The second option is to take action no matter how you're feeling—whether you're ready or not.

Which would you choose?

If you go with Option 1, you just signed up for a lifetime of waiting, stuck in limbo. When you struggle with Agency, this lack of trust in your abilities can affect you so deeply that it leads to a kind of rigidity or inflexibility you are at a loss to understand. You hesitate. You wait. You tell yourself you'll start when you have everything figured out. But lo and behold, you never feel ready—it's always "tomorrow." But tomorrow turns into *next week,* next week becomes *next year,* and before you know it, you've delayed yourself into oblivion.

On the other hand, if you pick Option 2, you've chosen a recipe for progress and success. Action—any action—creates momentum. Even if you stumble, you're learning, moving forward, and building trust in yourself. Like the lives of so many successful people we admire, like Zamani and his brother, people with Agency are able to act before they feel ready or despite not having it all figured out.

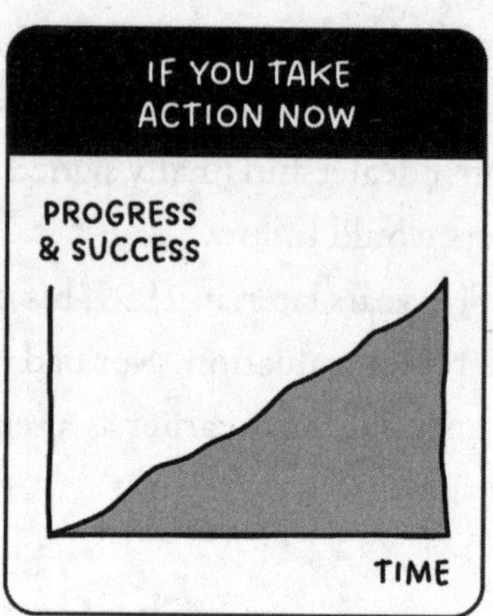

In fact, for many people, the worry that you don't have what it takes might not even be the real issue. The real culprit? The belief—or let's be honest, the excuse—that you need to *feel* confident before you're able to act (*When I feel ready, then I'll start*).

That's a lie. Confidence isn't a prerequisite for action—it's a *result* of it.[2] You have to *earn* confidence. And how do you earn it? By doing the work. If you want to do anything "with confidence," you have to start. Research backs this up. The steps are simple but annoyingly unavoidable:

1. Take action.
2. Build skills and develop your competence.
3. Boost your self-belief (hello, Agency).
4. Ride that sweet wave of earned confidence.

The word *confidence* comes from the Latin *con* (with) and *fidere* (to trust). So, confidence can be seen as an act of trust. We trust ourselves. Then that confidence makes it easier to take even *more* action. And suddenly, you've built a self-perpetuating cycle of growth.

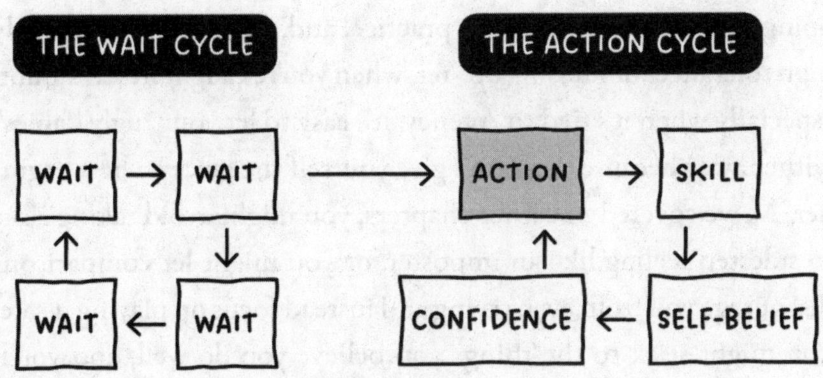

Lowering the Stakes

In her classic book *Drawing on the Right Side of the Brain,*[3] Betty Edwards shares this sweet anecdote from art teacher Howard Ikemoto. "When my daughter was about seven years old," he explains, "she asked me one day what I did at work. I told her that I worked at the college—that my job was to teach people how to draw. She stared back at me, incredulous, and said, "You mean, they forgot?"

That question cuts deep because it's true. Somewhere along the way, most of us *do* forget. We forget how to play, how to explore without fear, and how to trust our natural creativity. We trade our childlike curiosity for a need to be competent. And our measure of being competent is elusive. If we lack Agency, we almost never feel ready.

In his memoir, Pixar cofounder Ed Catmull writes that "early on, all of our movies suck . . ."[4] He calls these early mock-ups "ugly babies." He explains that every Pixar movie, no matter how brilliant it ends up, starts as a total mess. "They need nurturing—in the form of time and patience—in order to grow."

This is true for most things in life. Growth takes time. Developing skills requires patience, practice, and, yes, an uncomfortably high tolerance for messing up. Yet, when you're caught in self-doubt, especially when it's tied to Agency, it's easy to let your "ugly babies" wither away because you don't give yourself the grace to be a beginner. As we covered in earlier chapters, you might avoid taking risks to sidestep feeling like an imposter, or you might let comparisons discourage you from even trying and instead focus on playing it safe. You might stick to the things you believe you do well, and you'll find rationales to hold yourself back in all the areas where your insecurities lie. The problem is, when avoiding discomfort becomes your default, you end up saying no to growth without even realizing it. You miss out on the people, ideas, and opportunities that could have expanded your world—simply because you didn't give yourself the chance to try.

The self-doubt spurred by a lack of Agency keeps you stuck in inaction. It's like trying to drive with one foot on the brake checking your progress, and the other foot on the gas as you try to push, push, push. You don't go anywhere.

One of the participants in our online career development program, Juan Pedro, loves to paint—or at least he used to. He was struggling to get back into it. Every time he thought about picking up a paintbrush, he got stuck thinking about the end result. He wasn't painting to paint anymore—he was prejudging his work before it even existed. He was waiting to feel ready, to feel confident, to feel like he could guarantee a masterpiece before even starting. As you know, that's not how creativity—or life—works.

When Juan Pedro shared that he had two young daughters, his eyes lit up. You could tell he loved those kids more than anything. So, I challenged him: What if he approached painting the way his daughters approached play—with curiosity, experimentation, and zero concern for the end result? I wanted him to enjoy the act of painting itself and to find his passion in the act of "doing" rather than the "ending." I wanted him to keep his focus on the present moment.

Kids don't overthink. They're masters of "Let's try and see what happens!" They have unbridled enthusiasm and are not boxed in by the notion of "the right way" to do things. Even if they don't know how, they jump in, make a mess (sometimes a very big one), and figure it out as they go. They learn through play, not fear.

This power of play is why five-year-olds often outperform most adults in an exercise called the Marshmallow Challenge. In an experiment shared by Autodesk fellow Tom Wujec in his TED Talk,[5] he explains how small groups are given uncooked spaghetti, tape, string, and a single marshmallow. The goal is to build the tallest free-standing structure that will support the entire marshmallow on top in eighteen minutes. Adults tend to spend most of their time establishing a dominance hierarchy, strategizing, and consulting, and then turn to building with about six minutes left on the clock. The kids? They dive right in. They build, break apart, rebuild, and giggle when things go wrong. And as a result, they get better results than most adults.

Why? Because it's all play to them, a series of experiments to see what works and what doesn't. Five-year-olds adopt a "trial and learn" mindset, not a "trial and error" one. Every wobbly structure is just another chance to test, tweak, and try again. They see their "ugly babies" as opportunities for growth.

Juan Pedro took the play challenge to heart. He ditched the idea of creating masterpieces and started painting just for fun. The more he focused on play, the less pressure he felt—and the more spontaneous and creative his work became. He started nurturing his "ugly art babies," and in doing so, rediscovered the joy of creating.

PRACTICE: The Daily Dabble

Play is one of the most overlooked tools for better performance and a better mindset. It sparks creativity, yes. But it also reduces stress, boosts productivity, and helps you slip into that elusive state of flow (where peak performance feels effortless).[6] Not only that, it can strengthen your Agency and make you better at your job. In a study of more than 500 Nobel Prize–winning scientists, researchers found that they were almost three times more likely than the average scientist to regularly dabble in creative hobbies like painting, playing music, or writing. Even more striking is that they were twenty-two times more likely to be amateur performers, from actors and dancers to magicians.[7] Many of them credited these playful pursuits with helping them to think differently, spot connections others miss, and become better problem solvers.

So, here's your experiment:

- **Set aside just ten minutes a day for a creative, playful activity where the outcome doesn't matter.** Doodle. Play with clay. Strum a guitar. Write the worst poem ever written. The goal isn't to impress anyone—it's to remind yourself that it's okay to make a mess, to not be great, to explore without pressure. You're desensitizing yourself to the "ugly baby," rewiring your brain to embrace experimentation, and diversifying your interests.

- **At the end of the week, reflect.** Did it get easier to try without judging? Did you enjoy being a beginner? If yes, keep going. If no, try something new. Make play part of your self-trust practice.

Use Play to Find the "Right Fit"

One of the most overlooked benefits of play is that it helps you *explore*. It gives you permission to try things out—without judgment, without pressure, and without needing to know how it'll all turn out.

That spirit of experimentation is exactly what's missing when people feel stuck or lost in their careers. For example, many people become incredibly skilled at what they do, but they end up boxing themselves into a definition of success that starts to feel small, limiting, and constrained.

They might have task-specific Agency ("Another report? Easy."), but their generalized Agency is weak ("I'm not sure there's anything else I'm good at."). Amber, for example, had worked as a business analyst for more than a decade. She was excellent at her job, with stellar performance reviews, and she was known as the go-to person for anything related to financial modeling. But over time, she started to stagnate, feeling restless, frustrated, and uninspired.

Some researchers call this *boreout*[8]—feeling chronically bored and under-stimulated by your work that no longer feels stimulating, meaningful, or aligned with your growth. The irony was that Amber didn't even realize she was stuck because she hadn't given herself the chance to try anything else. Her job was no longer a fit, but she couldn't see it.

The right fit is not something you always know from the start, and it can change. Fayçal and I often speak about how tragic it would be if someone had an innate talent, yet never figured it out because they never tried. What if I have mad snowboarding skills but never find out because I've never bothered to strap on a board? (At least I found out I could hold my own as a Latin dancer and that I have a knack for putting together strategy pitch decks that influence executives.)

The point is, you don't discover what you're good at by sitting around thinking about it. As American journalist David Epstein writes in his book *Range,* "First act and then think. . . . We discover the possibilities by doing, by trying."[9] You might have a gift for storytelling, an aptitude for leadership, or the potential to be an amazing coder, but you'll never know unless you experiment. You have to actually try things to see where you learn fastest, where your natural talents lie, and what energizes you. This is how you gather the evidence that yes—you *do* have what it takes.

It's a key part of maximizing your advantages. Without real-world experience to draw from, your insight into yourself will be limited. Think of it like a buffet. You can't know which dish is your favorite unless you taste them all.

This same trial-and-learn approach applies to the big questions: "What's my passion?" or "What's my purpose?" The truth is, passion isn't found, it's built. It's active, not passive. It's through the act of trying, putting in the effort, gaining momentum and mastery, that your curiosity, interest, and passion start to grow. Confidence builds too. And as that confidence grows, so does your sense of Agency—the belief that you *can* do new things and grow into new roles.

This is one of the greatest gifts playfulness gives us: the permission to try. If I hadn't given myself permission to try, I would never have left the corporate world. I would never have stepped foot on a stage. I would never have coached a client, posted that first video on social media, or launched my business. I wouldn't have asked Fayçal to marry me (yes, I did that), I wouldn't have started a PhD, and I certainly wouldn't have written this book.

Ask yourself, *How can I really know how to do something* before *I've ever done it?* It's like being a parent. I've watched my brother and sister-in-law bring three beautiful kids into the world and doubt never stopped them. A friend recently said, "We have no idea what we're doing as parents . . ." Then she paused, smiled, and added, "And it's amazing!" Why don't we take that mindset into the rest of our lives?

The most successful (and happiest) people *believe* they can—and if they objectively lack the experience or skills they need, they're willing to ask for help, seek resources, or try a different approach. Life is about expansion—about stepping into new roles, learning through doing, and growing into the person you're meant to become. Play is the bridge that can help you get there.

PRACTICE: Create a "Play Practice" with Side Quests

In video games, there's usually a main quest: a central mission you need to accomplish to complete the game, using the tools you've already gathered as a player. But, in many games, the real secret to leveling up is through side quests. These optional challenges let you explore new areas, gather extra tools, and gain resources that make the main quest easier. To build your sense of Agency in real life, think about setting up your own side quests—ways to intentionally broaden your skill set and build your toolkit of resources. Take charge of your learning by creating your own "play practice" where you're free to take risks, experiment, and try new things. There are four steps: Choose, Play, Reflect, Adjust.

Step 1: Choose Your Adventure—Set Your Side Quest

Pick something low-risk and mildly exciting. It could be a creative project, a new responsibility at work, or a long-overdue attempt at something you've always said "Someday, maybe . . ." about.

Some examples of side quests:

- Come up with a new brainstorming method for your team.
- Write an article, LinkedIn post, or blog about a topic you're still figuring out.
- Sign up for a community project, like helping at a food bank or joining a neighborhood cleanup.

"Bored" business analyst Amber who was stuck in the Stagnation Zone set a low-risk side quest to volunteer with the marketing team to develop a financial model for a new product launch. Was it a huge leap? Not really. But it nudged her into a fresh environment and forced her to think on her feet.

Step 2: Play—Dive in and Experiment

This is where the fun (and sometimes mild panic) happens. Jump into your side quest with curiosity and start experimenting. Remember, it's low stakes, so there's no right or wrong outcome.

Amber treated her side quest like a mini-adventure. She concentrated on meeting new people, learned how a different team operated, and soaked up everything she could about a world outside her usual spreadsheets.

Step 3: Reflect on the Experience

After the side quest, take a moment to look back. How did it feel to step into something without feeling fully prepared? What was your mental state before, during, and after?

Amber was surprised by how energized she felt. She picked up some basic skills in market research, consumer behavior, and creative strategy, made a few mistakes (because that's part of the deal), and discovered she had a knack for translating numbers into compelling stories the marketing team could use in their campaign.

Step 4: Adjust and Refine

Now, tweak your process for the next quest. What worked? What didn't? What's your next experiment? Then, return to step 1 and do it all over again with a new side quest, but set a date.

Amber walked away from her side quest with a new outlook: "This experience has completely changed my perspective. Honestly, it's made work feel exciting again. I'm even considering a short-term placement within the marketing team and might transition there in the future."

Side quests take the pressure off getting it right and help expose you to new experiences without requiring meticulous planning. You're broadening your skills and building your toolkit, while learning how to apply what you've gained to new and unfamiliar challenges. And who knows? You might just uncover a hidden talent or develop a new passion along the way. Set your own side quest *today,* and give yourself ten days to complete it. No excuses.

• • • •

Action builds Agency, and Agency builds confidence. So stop waiting for certainty. Stop waiting to feel "ready." Do the thing—*even if you doubt yourself.* The only way to get where you want to go is to begin.

CHAPTER 13

The Gift of Inner Authority

Agency in the Here and Now

Building Agency doesn't come from waiting. It comes from action. On the one hand, you can interrupt the imposter thoughts, step out of the comparison trap, and *do the thing*—even if confidence hasn't shown up yet. And that's powerful. But there's also a deeper, more expansive path. It taps into something you already have—the gift of your inner authority. You just need to remember and honor it.

George, a doctoral student at Berkeley, rushed into his graduate statistics class late one day. On the board, two problems were written down. Thinking they were homework, he copied them down and then quickly turned his attention to the professor. Later on, when he sat down to tackle them, George realized these weren't your average textbook problems. They were harder than *anything* he'd tackled before. Days passed. He struggled, but he kept going. Finally, he cracked them. Satisfied, he turned in the assignment to his professor, apologizing for being late.

Except . . . these weren't homework problems. They were two of the most infamous unsolved theorems in statistics that had stumped mathematicians for years. George's professor was stunned.

Yet George, completely unaware that the problems were "impossible," solved them.[1]

This isn't some feel-good fictional anecdote I cooked up. This story is 100 percent true. George Bernard Dantzig went on to revolutionize the world of mathematics. He invented linear programming, which used a new approach to math to solve real-world problems. Thanks to his work, industries like logistics and transportation became wildly more efficient. Delivery routes, airline schedules, even supply chains were all streamlined because of him.

And that's the lesson. Lack of belief in ourselves holds us back. Imposter thoughts make us hesitate. Comparison makes us feel like we're behind. The fear of not being ready keeps us stuck. But Dantzig unknowingly sidestepped all of these. He didn't overanalyze whether he was "qualified" to solve those problems. He didn't compare himself to the great mathematicians who had failed before him. And he didn't wait until he felt certain he had the right answer—he just kept working.

This is the essence of Inner Authority, which is the antidote to inefficacy. It's that quiet, unshakable confidence that comes from trusting what you know and showing up fully, even when you don't know how things will turn out.

Honoring Your "Inner Authority"

This is the essence of honoring your Inner Authority:
Trust your own skills and capacity to learn, shift away from comparison and self-doubt, and focus on what you can do with confidence and authenticity.

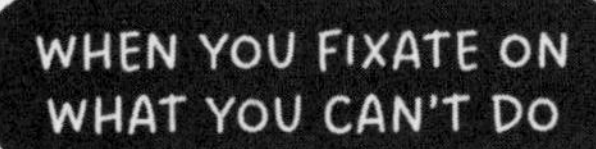

I CAN'T...
I DON'T KNOW
WHAT TO DO

WHEN YOU HONOR YOUR
INNER AUTHORITY

I HAVE TOOLS,
EXPERIENCE & INSTINCTS
TO DRAW ON

The real world doesn't make it easy to trust yourself. It constantly presents challenges that feel just beyond your current ability. In those moments, it's easy to look around, size up the competition, and assume that everyone else has it all figured out. And when that comparison takes over, doubt creeps in. You start to wonder, *Why try at all if I'm just going to fall short?*

Of course there will always be others who seem better equipped, more qualified, sharper, shinier, more *something.* The real difference between the people who do the hard things and the ones who don't isn't talent, and it isn't usually skill. It's *belief.* At the heart of Agency—and the Big Trust that grows from it—is your belief in your Inner Authority. It's the ability to come back to an unshakable trust in your own unique, individual strengths, even when self-doubt is doing its best to derail you.

As you know, I've been in that spot myself—caught in a loop of self-doubt that made me question my worth (from low self-acceptance) and second-guess my abilities (thanks to shaky Agency). In my first three months in banking after my big switch from corporate law, I fell straight into this trap. I was convinced I didn't belong. One day, I sat across from a mentor, Mel, and let my self-doubt come pouring out: "I don't feel like I have the skills to succeed here. I haven't mastered financial modeling. I'm terrible with Excel. What if someone finds out I CAN'T do this?"

Mel smiled. She let me ramble and then hit me with a truth I'll never forget: "Shadé, you're not here because of what you CAN'T do. You're here because of what you CAN do. Stop focusing on what you lack and start finding ways to bring your strengths to life."

Mel wasn't telling me to fake it or ignore my gaps. She was telling me to *stop obsessing* over them. To redirect my focus to what I *did* bring to the table. Qualities like my enthusiasm, my passion for solving problems, my curiosity, and my ability to learn fast. That conversation handed me the keys to my Inner Authority—the permission to trust myself and lean into the essence that was already there.

That conversation marked a turning point. It kept me from walking away from the industry when self-doubt was telling me to run. I stuck it out. I persisted. I learned. I built skills. And in the process, I began to trust myself more. That's the gift of honoring your Inner

Authority. It lets you see past the noise of your self-doubt and focus on the qualities you already have, right here, right now, knowing that they will grow.

Remembering What You Bring to the Table

Inner Authority doesn't give you new superpowers. It doesn't magically *create* abilities, skills, or talents out of thin air. What it *does* is unlock the potential that's already within you. The fact is, most of us are far more competent, stronger, wiser, and more capable than we give ourselves credit for. The problem, though, is we're so busy fixating on what we *can't do* that we never tap into what we *can do.*

Here's what Inner Authority could look like for you:

- You walk into an investor pitch meeting. The other founders have fancy degrees, glossy decks, and polished delivery. You feel out of place. **Your Inner Authority reminds you:** Your power isn't in mimicking their polish. It's in your raw, no-nonsense conviction. Speak from that place.
- You're scrolling social media and see your competitor's product launch getting tons of buzz. Their branding looks flawless, their following massive. Self-doubt creeps in. **Your Inner Authority reminds you:** Your creativity doesn't live in comparison. Your ideas are bold, your vision is unique. Stay focused on what you do well and what differentiates your brand.
- You're in a meeting. The room is buzzing with technical jargon you don't fully grasp. The panic sets in. **Your Inner Authority reminds you:** Apply your strength of curiosity. Connect the dots in ways others don't see. You don't need to pretend you

know it all. Trust the part of you that's always been willing to learn and brave enough to ask.

When you trust your skills and your ability to figure things out, challenges don't feel insurmountable. You're not bogged down by the enormity of the task or paralyzed by *What if it's too hard?* Instead, you're locked in on *what's possible*. In that space, self-doubt doesn't stand a chance.

But let's be real. Most of us don't live in that zone too often. What's more common is that we let self-doubt creep in, looping through our minds all the ways we're not good enough. It amplifies every flaw, every gap, every insecurity, until we're convinced we shouldn't even try. The challenge feels too big. The skills feel out of reach. Someone else must be better suited for this. And just like that, we shrink.

When you connect to your Inner Authority, self-doubt quiets and self-trust begins to take its place. Someone else's success doesn't feel threatening; it feels inspiring. You don't see their achievements as proof you're falling behind; you see them as evidence of what's possible. That's the heart of Inner Authority—staying anchored in your own path and your own progress, and showing up as your fullest self.

Honoring your Inner Authority doesn't guarantee every day will feel like a win. Some days are messy. But it gives you the courage to show up anyway. It's what lets you follow Richard Branson's advice: "If somebody offers you an amazing opportunity but you are not sure you can do it, say yes—then learn how to do it later!"[2]

When you honor your Inner Authority, you affirm that "I am open for opportunities." You let yourself, God, Jehovah, Allah, the Creator, the Universe, the Great Cosmic Force—whatever you connect with—know that you're *ready*. Ready to act. Ready to grow. Ready to take on what's next. You may not have all the answers yet, but you trust that you'll find them along the way.

TAKE A MOMENT:

Imagine what you could achieve if you fully embraced your Agency and honored your Inner Authority today, this week, and beyond.

The Third Attribute

AUTONOMY

CHAPTER 14

Do My Choices Matter?

The Core Question of Autonomy

Bruno, an early executive coaching client of mine, was a force. He was charming, charismatic, and quick to make decisions. Yet for all his brilliance, he had an uncanny ability to find fault in *everything.*

In our very first session, he walked in, impeccably dressed in a perfectly tailored suit, and as I shook his hand to welcome him, he immediately launched into a litany of frustrations: "The traffic was brutal. The roadwork's a nightmare. I tried to grab a coffee before the session, but between the delays and the insane line at the café, I had to skip it. Wow, why is it so hot in here?"

He hadn't even sat down yet.

Bruno was the founder of a fast-scaling tech company in the hospitality industry. He'd reached out because he was having trouble navigating his company's expansion. It should have been an exciting chapter, yet he couldn't stop complaining.

"These new hires don't get what our business is about. And my executive team? Half the time they make my job harder when they should be making me look good in front of investors." Before I could get a word in, he was already onto his next grievance.

As he vented, I noticed a theme. Everything was happening *to* him, and everything felt out of reach.

I gently interrupted, "Sounds like you've got a lot on your plate." I could only imagine how heavy his storm of frustration must feel.

"Of course I do," he shot back. "There always is. I feel like I'm constantly fighting against things I can't control. It's exhausting. It shouldn't be this hard."

Now he'd caught my attention. "What makes you say that, that it shouldn't be this hard?" I asked.

Bruno let out a deep, exasperated sigh. "Because it feels like everything is working against me. It's like no matter what I do, there's always something or someone messing things up. I just think, after all this work, it should be easier to keep things on track. Sometimes I wonder if any of it even matters. Maybe it's all pointless."

I could hear the sadness in his voice. Bruno wasn't just overwhelmed by immediate business challenges—he could get a handle on those. He was grappling with something deeper: the creeping doubt that all his efforts didn't matter. He was feeling powerless, questioning whether he could actually influence the course of his life and work—or if he was just along for the ride.

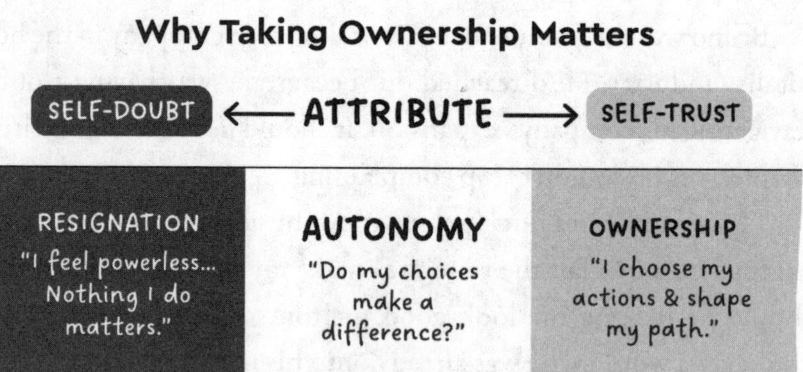

The defining challenge of Autonomy, the third Attribute in our Doubt Profile, is reclaiming your sense of ownership over your own choices. At our core, we all crave the ability to shape our own lives.[1] We want to feel like we have a say in what happens next. When Autonomy is a vulnerability, you live in a state of resignation. While it shares similarities with a lack of Agency (which is about doubting your capacity), the struggle for Autonomy is about doubting whether your efforts matter. You feel like life is happening *to* you rather than something you can actively shape.

Of those we surveyed, 40 percent didn't believe their actions made a difference. Another 20 percent were on the fence—neither agreeing nor disagreeing that they had control. That's three in five people either doubting or uncertain about their ability to influence their own outcomes.

If you struggle with Autonomy, you start to believe that external forces dictate your success more than your own actions. You hesitate, not because you lack the desire or drive, but because deep down, you don't believe that actions you take will make a difference. Why speak up if it won't change anything? Why apply for that role if the odds are already stacked against you? Even if you have high levels of Acceptance or Agency, it won't matter if you don't believe you have the right or power to act in the first place. The result is resignation. When you don't trust yourself or your ability to shape your circumstances, it's easier to let go of the fight than to risk failing in it.

This is what I was seeing with Bruno. He had a sharp mind and a solid track record. But like many people under pressure, he became stuck seeing the world through one narrow lens: Everything was *outside* his control. The traffic derailed his entire morning, the weather was unbearable, the new hires didn't "get it," and his exec team kept missing the mark. Even his investors weren't "on the ball anymore."

And yet, here he was—running a successful company, surrounded by a team that had helped him build something meaningful. But none of that registered.

This happens more than you'd think. When Autonomy dips—especially in stressful times—you stop believing you have influence. Instead of stepping forward and taking ownership, you freeze or retreat. You sink deeper into self-doubt, replaying the stories that keep you stuck—watching life happen to you instead of stepping up to change it. To cope, you might fall back on the familiar habits of complaining, resenting, or dwelling.

1. The Complaining Habit. Venting feels good in the moment. It's a pressure release, a way to seek validation and express frustration.[2] But research shows that when you complain, you're actually *reliving* the negative experience in your mind, replaying it over and over in vivid detail.[3] Every time you rehash that terrible meeting or that rude barista, your brain paints an even *more* vivid picture of just how terrible it was.

Psychologist Travis Bradberry warns that "repeated complaining rewires your brain to make future complaining more likely."[4] The more you do it, the easier it becomes—until you've become a negativity magnet. And this makes you feel powerless.

Worse still, it spreads. Negativity is contagious. Before you know it, you've built a toxic echo chamber[5] where frustration bounces between you and others, growing louder with every pass. Instead of breaking free from that powerless feeling, you reinforce it. And ironically, it often pushes away the people who could support you most, so you feel even more isolated.[6]

2. The Resentment Habit. Resentment is a slow-acting fuel for self-doubt. Maybe it's a promotion you didn't get or a colleague getting recognition you feel they didn't deserve. Your mind zooms in on every supposed advantage they had—*they went to a fancier college . . . they only got hired because they checked a diversity box.* And just like that, you've convinced yourself that everything good is happening out there for them, not you.

This mindset, called *other-enhancement,*[7] hooks you into thinking life is rigged against you. Though it might cushion your ego for a while, you're so busy replaying how unfair things are that you forget you still have the power to change something.[8]

My grandmother used to say in Farsi, "Khoon-e to masmoom nakon"—"Don't poison your blood." But resentment does exactly that. It clouds your thinking, drains your energy, and leaves you so focused on other people's wins that you miss your own open doors.

3. The Dwell Habit. As we learned in chapter 1, your brain craves certainty. And sometimes the past—no matter how painful—feels more certain than the unknown. That's how dwelling starts. You rehash old hurts. You replay old conversations in your head. You relive moments you wish had gone differently. You convince yourself that you're just "processing" it—but what you're really doing is rehearsing helplessness.

Dwelling is an attempt to *make sense* of why things haven't gone your way. It creates a story: that the game is rigged, that no matter what you do, it won't change anything. So instead of acting, you overanalyze. Instead of moving forward, you stay stuck in what's already happened.

It's not that the pain wasn't real. The betrayal, the missed opportunity, the moment that still makes your stomach sink—those things matter, and they sting. But when you dwell, you let them become *more* powerful than they need to be. You build your present—and sometimes your future—around them. Slowly, you stop looking for what could be different because you're too busy replaying what wasn't fair. You're left thinking, *Why does this always happen to me?*

When your sense of Autonomy is low, life starts feeling like one long conspiracy against you. And your brain (or more specifically, your Gatekeeper filtering the information you receive) is desperate to confirm that belief. It tunes out anything that contradicts it, cherry-picking every struggle and cementing the "I'm powerless" narrative.

Of course, there's a world of difference between a "Why me?" mindset and living through genuine injustice, systemic oppression, or the many places and situations where Autonomy is practically nonexistent. Those facing genuine powerlessness need real solutions and support, not a pep talk.

But for everyone and every situation where we *do* have choices—about where we live, whom we work for, how we respond to life's inevitable curveballs—our barriers are often largely *self-imposed.*

The truth is, you're not powerless. Even when you can't *change* the situation itself, you have the power to *change* how you experience it and how it shapes your future.

We All Have a Locus of Control

Whether you realize it or not, you carry a belief about how much influence you have over your life. Psychology researchers call this your *locus of control.*[9] The word *locus* comes from the Latin *loci* which means "place" or "location." In this context, it refers to where you place your control—inside yourself or in the hands of external forces. Your locus of control is shaped over time, by your early experiences, the messages you observed from the people around you, and the culture you grew up in.[10]

- When you believe that what happens to you is a result of your choices, you have an *internal locus of control*. You believe your actions shape your future, and you focus on what you can influence—like your effort, preparation, and response to challenges. This belief reflects the foundation of a healthy sense of Autonomy.
- When you believe your life is mostly shaped by luck, circumstances, or other people's decisions, that's an *external locus of control*. You focus on what you cannot influence. You believe life happens *to* you, and when things go wrong, you tend to feel powerless.

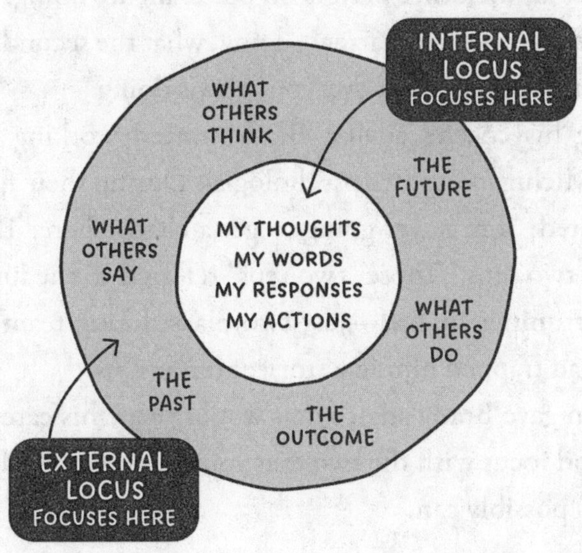

Think back to your school days. If you aced a test and thought, *That's because I studied,* or failed and thought, *I didn't prepare,* you were learning that effort leads to results. But if you worked hard and still got overlooked by biased teachers, or if you were constantly undermined by bullies or unsupportive adults, maybe you started thinking, *Why bother? Nothing I do changes anything.* That's one way a sense of powerlessness can take root.

Shifting into a state of powerlessness isn't always obvious. It can creep into our psyche undetected by us, and sometimes it takes another person to point it out.

Quarterback Tom Brady, the NFL legend, faced this early in his career. In his second year playing college football with the Michigan Wolverines, Brady was stuck on the bench. He barely played and felt overlooked. "I would complain all the time," he admitted,[11] constantly ruminating on how unfair it felt, how others got more field time, more opportunities, more attention.

His coach eventually called him out: "Brady, stop worrying about what all the other players on our team are doing. All you do is worry about what the starter is doing, what the second guy on the depth chart is doing, what everyone else is doing."

Taking his coach's advice, Brady started working with Greg Harden, Michigan's sports psychologist. During their first session, Brady vented: "I'm never going to get a chance here. They're only giving me two reps." Those "two reps" referred to the limited practice opportunities he had—just two plays during team drills. His mindset had trapped him in a story of unfairness.

Harden gave Brady advice that would shape his career. "Just go in there and focus with the two that you got and make them as perfect as you possibly can."

Brady realized that while he couldn't control how many reps he was given, he could control how he showed up for them. So he did just that. Instead of focusing on what others in the team were doing or how "wronged" he felt, he focused on what he could control. He poured his enthusiasm and his energy into those two opportunities. Over time, the two reps became four, then ten. Brady said, "With this new mindset that Greg instilled in me—to focus on what you can control, to focus on what you're getting, not what anyone else is getting, to treat every rep like it's the Super Bowl—eventually, I became the starter."

This shift set him firmly on his path to becoming one of the greatest quarterbacks of all time.[12]

What changed? Not the reps, not the politics. It was that his locus of control shifted inward. He stopped giving energy to what he couldn't change and took full ownership of what he could.

Yes, there are real barriers in the world—class, gender, race, health, ability, discrimination. Autonomy doesn't mean pretending those don't exist. But in nearly every situation, there is *something* you can shape—perhaps most importantly, how you feel about yourself. Is your self-doubt defeating you? Are you stuck in a story that your choices and actions don't matter?

You don't need a big plan or a major breakthrough to start changing your sense of control. In fact, it often starts with something small: the language you use to describe your situation. The words you speak—out loud or in your head—shape how you see yourself and what you believe is possible. Low-autonomy thinking often sounds like: "I have to" or "I can't." These phrases quietly reinforce the idea that you have no choice—that life is just happening *to* you.

Replace those phrases with "I choose to," "I want to," or "I'm committing to." Studies show that subtle changes in language can boost motivation and restore your sense of control.[13]

"I *have to* get up early" becomes "I choose to." "I *have to* finish this paper by 5 p.m." becomes "I *commit* to wrapping up by 5 p.m." Even in difficult moments, choice exists. For instance, if you're grieving and not ready to get out of bed, "I *can't* get out of bed" becomes "I'm *choosing* to rest." This habit of intentional language helps you take ownership of your actions.

The Promise of Taking Ownership

Taking ownership of how you show up is where your power begins. It changes the focus from what isn't controllable to what is.

Instead of wasting energy on office politics, market crashes, ways you've been hurt, or someone else's bad decisions, you center your energy on what you *can* influence—your actions, mindset, choices, and next steps. As Viktor Frankl, who endured unimaginable hardship in concentration camps during World War II, wrote, "When we are no longer able to change a situation . . . we are challenged to change ourselves."[14]

Research shows that people who focus on what they *can* control report lower stress, stay more motivated, and perform better[15]—not because life goes easier on them, but because they stop fighting battles they can't win. They use their energy where it counts.

One of the most powerful benefits of Autonomy is that if something in your life isn't working, you don't just sit with the frustration, complain, or spiral in self-doubt—you do something about it. And when you can't change the situation? You change how you relate to it.

When you know you always have a choice, you *feel* powerful. You feel more in control of your life. And when you have that, it becomes a powerful buffer against self-doubt and a key part of building Big Trust. It's how you take back your power.

PRACTICE: Assess How Autonomy Is Showing Up for You

When you understand the role Autonomy plays in your Doubt Profile, you start to see where you've been giving away your power—whether to people, circumstances, or old stories about who you are. But you also begin to notice where choice still lives. With that awareness, you can start making small, consistent shifts.

Look back at your Autonomy score from the Doubt Profile diagnostic. What was it?

Autonomy Score: ________ Zone: ______________________

(For Zone, write in "Red Alert," "Hindrance," "So-So," "Hidden Strength," or "Superpower.")

Take some time to reflect on the questions below. You might like to journal your responses:

What are your vulnerabilities around the search for Autonomy?

- **Think back to childhood—how much freedom did you have to make decisions?** Did you feel like your choices mattered, or were decisions often made for you? How has this shaped the way you approach control and responsibility today?
- **How does self-doubt show up when you feel powerless?** Do you blame external circumstances for holding you back? Do you find yourself stuck in resentment, avoidance, or waiting for permission or perfect conditions before taking action?

- **When was the last time you reclaimed control in an uncertain situation?** How did focusing on what you could influence help you shift out of helplessness and take a step forward?
- **How might your life look and feel different if you fully believed you could shape your outcomes—even when external forces aren't ideal?** What would you stop postponing? What would you start creating?

When Autonomy feels out of reach, what strength can you lean on?

- If you're feeling stuck in powerlessness, draw on your **Acceptance**. What limiting belief about yourself might be holding you back from trusting your ability to create change? How could self-acceptance help you release that?
- If you're waiting for the "right" time to act, lean into **Agency**. What skills, experience, or personal qualities can you draw on right now to take action and remind yourself that progress comes from movement—not waiting?
- If emotions like frustration or discouragement are clouding your ability to act, turn to **Adaptability**. What emotional regulation strategy could help you clear your mind and focus on what you *can* control in this moment?

• • • •

In the pages ahead, we'll unpack how the erosion of Autonomy shrinks your world—keeping you planted in a "pot" that's too small, too safe, and ultimately too limiting. We'll explore how when life throws you hard things (and it will), having an internal sense of control can give you a new and powerful way to navigate the storm. Then, in chapter 17, we'll step back and introduce a deeper, perspective-shifting force—something that expands your sense of possibility and helps you reconnect with your capacity to influence what comes next. It's a shift that strengthens your autonomy and helps to lay the groundwork for Big Trust (the deep, grounded knowing that you can shape your future).

CHAPTER 15

Why Can't Life Be Easier?

Choose Your "Hard"

At age thirty-nine, after fifteen years as a middle school science teacher, Sonya felt the itch for something new. Urban planning and sustainable design had always intrigued her, given her love of studying how cities were built and how thoughtful designs made neighborhoods more livable. She'd daydreamed about making a career switch for years. But every time she thought about making the leap, she hit a wall of doubt and fear.

Her daily routine was predictable—wake up, teach, go home, repeat. It was safe, familiar, *comfortable.* The idea of changing careers felt terrifying, because it meant stepping into the unknown—and walking away from something she'd spent so many years building. Part of what held her back was the feeling that leaving teaching behind would somehow "waste" all the time and effort she'd already invested. That's the sunk-cost fallacy in action, when past investment starts dictating future decisions, even if it's no longer serving you. But for Sonya, there was also something deeper that kept her stuck in *comfortable*—the belief that she wasn't capable of managing the uncertainty that came with change.

And here's the irony. Her comfort zone wasn't actually comfortable. It was laced with resentment. She'd catch herself grumbling about her job, pointing fingers at the education system for why she felt so stuck, or envying colleagues who seemed perfectly content. This kept her tethered to her comfort zone, where it *felt easier* to stay put than to confront the uncomfortable truth that making a change was *her* responsibility. And let's be honest—it's no favor to her students for her to stay in the classroom while simmering with resentment, unable to give them the enthusiasm and energy they deserve.

Sonya's comfort zone didn't just impact her work life. She stuck to the same social circles and shied away from new experiences. The thought of meeting new people in the urban planning world or testing the waters by attending industry events was overwhelming. But each time she retreated, her self-doubt tightened its grip, leaving her feeling even less capable of steering her life.

The High Costs of Comfort

The reason why we prefer comfort over growth is because, like so many things, our brain is wired to be that way. We all have a psychological and emotional *safety zone,* where we think and act in ways that minimize risk and stress.[1] It nudges us toward "easy" over "hard," "familiar" over "unknown." It's our comfort zone. We feel safe in it. Protected.

Your comfort zone might look like clinging to the safety of a familiar routine. It's dodging new challenges—like taking on that big project or pursuing a creative opportunity—because learning new skills or working with unfamiliar people feels overwhelming. It's choosing a Netflix binge over having that tough conversation with your partner, or staying in a job you've outgrown because the unknown feels scarier than the dissatisfaction you're stuck with.

Whatever your own comfort zone looks like, know that you pay a huge price for it.

If you don't develop ways to tolerate the pain and discomfort that come with new experiences, you limit your ability to fully live. If you're a writer afraid of rejection, you never share your work and miss out on the joy of connecting with readers. If you're a manager but can't make tough decisions, you'll never test your ability to lead others.

The worst part about your comfort zone is that it silences you. Sonya had ambitious dreams. But as long as she stayed stuck, they remained just that—dreams.

Self-doubt and the comfort zone work together as partners in crime. They convince you that taking action won't make a difference, so you don't bother trying. But without embracing discomfort, you won't grow, and you'll never know what you're truly capable of.

YOUR COMFORT ZONE KEEPS YOU SMALL

WHERE YOU DISCOVER WHAT YOU'RE CAPABLE OF

A byproduct of our desire to stay safe in our comfort zone is that it can come with an unexpected side effect: boredom. Not the "I'm bored, so I'll scroll Instagram" kind of boredom—this is deeper. It's that restless, "blah" feeling, the ennui, where no matter how much you think about what you *could* do, nothing actually feels worth doing. Sociologist Corey Keyes introduced the concept of *languishing* as a neglected middle ground between depression and flourishing—when you're not mentally unwell, but you're definitely not thriving either. The idea went viral when Adam Grant spotlighted it in his 2021 *New York Times* article "There's a Name for the Blah You're Feeling: It's Called Languishing."[2]

When you feel like you have zero control over your life's outcomes, you fall into the *why bother?* mindset—that's the mindset of languishing.

Sonya was knee-deep in it. When I asked her what she thought was most important for motivation, she said what most people do: "I suppose purpose, knowing your 'why,' and having clear goals." Except, the science doesn't agree.

Yes You Can . . . Do the Hard Thing

Of all factors that have been studied to understand motivation, the most potent one is simply feeling like you're making progress.[3] Let that sink in. Progress keeps our motivation levels high, not purpose, because progress gives you a sense of control. There's the direct impact of your actions. You can see it. You can track it. You can feel it. And *that* feeling of control is what helps reinforce a sense of purpose and direction. When we're stuck languishing in our comfort zone, progress screeches to a halt. Instead of taking action, you start fixating on external factors and convincing yourself, *It's out of my hands. Nothing will change.*

When self-doubt leaves us feeling stuck, it's tempting to avoid the hard stuff. But this avoidance is a defense mechanism that reinforces cycles of inaction.

If you will bear with me, there's a lesson we can draw from bison and cows. Though these animals are closely related, their approaches to weathering storms—real ones—are different.[4] Cows are known to huddle together and walk away from a storm. Bison, on the other hand, have been observed to do something counterintuitive. They're known to face a storm. When cows walk with the wind of the storm, they end up receiving more of the storm. Bison face *into* the storm, which often gets them through the bad weather more quickly.

Though you might prefer to think of yourself as a friendly labrador or an agile gazelle, this image of cows and bison gives us a metaphor for how we approach discomfort in our lives.

It's *easier* to complain about the problem than to look for a solution. It's *easier* to resent the people who seem to have it easier, or blame the universe for letting us down, than to take responsibility. It's *easier* to cling to the comfort of what we know than to risk falling on our face by trying something new. You avoid taking action, so you don't see results. You don't see results, so you start believing you're powerless to change anything. And before you know it, you've convinced yourself that life is rigged against you.

One of the greatest belief-shifting statements I wish I learned earlier in life is this: "You get to *choose* your hard." It is key to making progress and reclaiming control. A Reddit post from 2021 captured this idea perfectly:

> Marriage is hard. Divorce is hard. *Choose your hard.*
>
> Being in debt is hard. Being financially disciplined is hard. *Choose your hard.*

> Communicating is hard. Not communicating is hard. *Choose your hard.*
>
> Life will never be easy. . . . But we can *choose our hard.* Pick wisely.[5]

While on the one hand the message oversimplifies things and is lacking nuance, on the other hand it gets the point across: "Hard" is often unavoidable. The type of "hard" you face, however, is to some extent up to you.

Doing something might be "hard." But doing nothing might also be "hard." It's a matter of perspective.

As James Clear writes in *Atomic Habits,* "Every action you take is a vote for the type of person you wish to become."[6] The small, deliberate actions you take help you "prove" to yourself that you *can* influence your own outcomes. And this reinforces your Autonomy.

For Sonya, starting over would be undeniably "hard." She'd have to juggle work with graduate school, rebuild her network in a new industry, take on financial risk, and wrestle with the inner voice whispering, *Who do you think you are to make this switch?*

But staying stuck was also hard.

One night, as she sat grading yet another stack of papers and her red marker started running out of ink, the realization finally sunk in. Both paths were hard, but one offered growth and excitement and the other promised more of the same, which would be even harder to live with over time.

In fact, the "easier" hard often comes with a higher cost later. Call it the "Comfort Zone Tax." You convince yourself to stick with the familiar, and you call it "practical." You're not really avoiding "hard"—you're just choosing a version of it that keeps you small. It's easier, at least in the short term. However, the cow mentality eventually catches up and you have to face the brunt of the storm. But it doesn't have to be that way.

The bison mindset faces the storm. It's the moment you decide to stop playing it safe and start backing yourself.

That's what Sonya did. She enrolled in a part-time online master's degree in sustainable design and took evening classes. Four years later, she graduated and landed a role at a sustainability-focused urban planning firm. What gave her the courage? She drew strength from her other Attributes. She leaned on Acceptance—knowing she deserved a more fulfilling path—and she tapped in to her Agency, trusting that she could learn new skills and handle the challenge. Taking those initial steps allowed her to reclaim her Autonomy and take ownership of her direction. "If I'd stayed comfortable in teaching, I'd be completely miserable, and probably very bitter too," she told me. "I would've known deep down I'd settled for less because I thought it would be easier. I've found strength I didn't even know I had."

To be clear, teaching wasn't "less." It just wasn't right for *her*. Someone else might leave urban planning to become a teacher—and that would be their leap into alignment.

This isn't about careers. It's about choices. The real win is having the courage to challenge yourself and pursue what makes you come alive. Your version of "more" is yours alone, and finding it means being brave enough to let go of comfort and walk toward what truly matters. It also means looking for and making your own luck.

PRACTICE: The Cost Reframe

When you're hesitating in the moment, your brain hyperfocuses on the *short-term discomfort* of taking action—the uncertainty, effort, or chance of failure[7]—and overlooks the long-term cost of doing nothing. It's called *temporal discounting*: We trade future costs for comfort now. Here's how you can flip that mental script *in the moment* of avoidance:

First, notice the hesitation—for example, you're staring at the application, the gym bag, the calendar invite you want to avoid . . .

Then, reframe by asking yourself:

- "What do I gain if I do this now?"
- "What will this cost me if I avoid it now?"
- "What will this cost me in six months? In twelve months?"

Finally, do something. Even if it's one tiny step. Submit the application. Put on the shoes. Schedule the call. By shifting focus to why it matters (the cost of *not* doing it), you disrupt your brain's default bias and create a pathway toward action.

Create Your Own Luck

Stepping outside your comfort zone doesn't mean you become immunized to pain. It means you get better at handling it. It doesn't guarantee that everything will work out, only that you have a greater *willingness* to expose yourself to the unknown and push beyond your current limits. It's the courage to act in spite of self-doubt and to *try*—the readiness to accept the inevitable friction and frustration that show up whenever you do something new or unpredictable. Over time, this habit of choosing "hard" teaches you to trust your own choices.[8]

When you accept that you can't control everything, paradoxically, you gain more control. You start trusting that you'll be able to handle whatever comes your way, even the parts you didn't plan for. And that is a powerful way to feel.

Of course, randomness plays a role in every life. A study published in 2023 found that 60 percent of managers had a defining career moment shaped by chance—being in the right (or wrong) place at the right (or wrong) time.[9]

But Autonomy gives you an edge. It increases your odds of experiencing the right kind of chance, or luck—the kind that moves you forward. Tech entrepreneur Jason Roberts calls it your "Luck Surface Area."[10] The more you show up, speak up, take risks, and put yourself out there, the more chances you create for luck to find you. Think of it as *earned good luck*. It's not handed to you. It finds you because you kept showing up. You made yourself visible. You made yourself *discoverable to opportunity*.

That's how multiple award-winning screenwriter and director Christopher Nolan approaches filmmaking. He has a reputation for being absurdly "lucky" with the weather. In shooting movies like *Inception, Dunkirk,* or *Oppenheimer,* he insists, "It's completely untrue. I'm very *unlucky* with the weather. But I made a decision early on that whatever the weather is, I will shoot. . . . We just shoot, whether it's pouring rain or the sun is out. And beautiful things can come from that."[11]

There's an iconic scene in *Oppenheimer* when the crew was filming a re-creation of the first test of a nuclear detonation. What "really made the sequence come to life," Nolan says, is that "this big storm rolls in with tremendous drama."[12] That storm was not planned, yet it made the sequence unforgettable.

Nolan didn't *wait* for cinematic magic. He showed up anyway.

That's the first part of creating your own luck: You have to show up consistently, without a promise of how things will go. That expands your Luck Surface Area.

But there's a second important piece to this. Nolan didn't make the most of that storm by chance. He was ready for it. And he was ready because he'd practiced the habit of showing up—again and again—through discomfort and unpredictability. He doesn't film in unpredictable conditions because it's glamorous. He does it because it's how he and his team have trained. They've built their capacity for discomfort in small, consistent ways by exposing themselves to unpredictable weather conditions on a filmset so that when opportunity strikes, they're ready.

You can do this in your own way, too, by leaning into your Autonomy. It starts by deliberately exposing yourself to small doses of "hard"—the everyday frictions that stretch you, train you, and slowly raise your threshold. Because, if you don't seek out even small doses of "hard," you don't grow. If you always deadlift the same 135 pounds at the gym, your muscles don't get stronger. If you only practice one scale on the harmonica, you won't improve. If you avoid hard projects, tough conversations, or unfamiliar territory, your confidence stalls. And if you avoid discomfort altogether? You shrink your Luck Surface Area.

Every time you *choose* to do something "hard," uncomfortable, or inconvenient, you're sending yourself a signal: *I trust myself to act, even when it's hard.* Each small decision becomes proof: *This was hard, but I chose to do it anyway. I am in control.*

Feel the discomfort, own it, and reinforce your Autonomy. That's how you create your own luck and pull off the burrs of self-doubt.

PRACTICE: Be a Bison: Microdose Your Hard

The best way to get better at doing hard things . . . is to keep doing hard things. But not all at once. Over time, repeated exposure allows your neural pathways to rewire.[13] You create a new baseline where you recover faster and adapt quicker than you thought possible. It's called *systematic desensitization*.[14] The key is to start with the smallest, most manageable dose of the "hard thing": a microdose. Here are five steps to help you microdose your hard and retrain your brain's relationship with discomfort—just like a bison in a storm.

Step 1: Identify Your Hard.

What's something you've been avoiding because it feels too uncomfortable, intimidating, or overwhelming? Initiating conversations? Pitching your idea? Posting your work online? Speaking in front of others? Write it down.

Step 2: Break It Down.

Shrink the task into the tiniest, least-threatening version possible. These are your microdoses. If your "hard" is initiating conversations, maybe your first microdose is making eye contact and smiling. If it's public speaking, maybe it's sharing a voice note with a friend. If you're building visibility online, maybe it's commenting on one post a day. The smaller, the better. Your goal is to make the first steps so easy, you *can't* say no.

Step 3: Schedule the Dose.

Pick your rhythm. Daily? Maybe three times a week? Keep it consistent. Remember, bison don't sprint through storms; they face them step by step.

Step Four: Track the Shift.

Note how you felt before, during, and after each microdose. At the end of each week, take a moment to reflect on your experiences. You're creating proof. Every experience of "hard" reinforces that you can handle more than you think.

Step Five: Scale Up When Ready.

Once a microdose no longer feels challenging, raise the bar. Say "hi" and ask a question. Post your work and invite feedback. Gradually amplifying the discomfort makes sure you're expanding your tolerance in a sustainable way.

The goal isn't to eliminate discomfort entirely but to build your capacity to handle it. You're expanding your comfort zone and, at the same time, expanding your Luck Surface Area. You're entering new rooms, meeting new people, sharing new ideas, expanding your skill set. And with each step, you're giving *earned luck* more opportunity to land. Before you know it, you'll have more confidence, a stronger sense of Autonomy, and less time wasted on avoiding the things that used to scare you.

This is Big Trust in motion. Not the absence of fear, but the decision to move forward anyway, because you trust your ability to handle whatever comes. You can do hard things.

• • • •

When you choose to seek out challenges that push you beyond your limits, that force you to adapt and evolve, you show yourself what you're capable of. So, be a bison: Face the storm, focus on what you can control, and choose your hard. Take action and watch what happens when you stop avoiding and start owning your Autonomy.

CHAPTER 16

Why Is This Happening to Me?

When "Hard" Chooses You

As a teenager, Fayçal was an unstoppable force. A regionally ranked triathlete, he woke up at 4:30 a.m. on weekdays to train before school. After classes, he pitched in at his family's small stationery, newspaper, and magazine store, serving customers, stocking shelves, and cleaning up after hours, as well as on weekends. Academically, he was just as driven, earning a spot at one of Australia's top universities. Fueled by a passion to make a difference, he took a year off from college to do volunteer community service on the remote island of Pohnpei in Micronesia. At nineteen, he felt invincible. But just a few months into volunteering, he contracted a near-fatal staph infection. Well-meaning but inexperienced local doctors prescribed months of broad-spectrum antibiotics. They obliterated his gut microbiome. He lost twenty-two pounds as well as his hair—and with that, a piece of his identity.

This marked the beginning of a nearly thirty-year battle against a mysterious illness that no doctor fully understood. Chronically fatigued, plagued with gastrointestinal problems and frequent and random brain fog, the bad days were overwhelming. "Here I was," he says, "once this top athlete and a dedicated, hard-working kid, and now I couldn't even rely on myself. I had every reason to doubt myself and what I was capable of. One week I'd be fine; the following I couldn't get out of bed. And since doctors weren't able to diagnose the core issue, I started doubting everything—my resilience, my strength, even if there was something wrong with me or if it was all in my head."

Fayçal had spent his life believing that if he worked hard enough, he could control his future. But what happens when that belief is shattered? When life doesn't go according to plan, no matter how disciplined or driven you are? The hard reality is that you don't get a say in the hand you're given. You don't get to choose where you start, what obstacles come your way, or when life will knock the wind out of you. Hardship doesn't discriminate.

Maybe you grew up in a family that didn't believe in you. Or, just as you were gaining momentum, a devastating setback struck. Maybe betrayal, illness, loss, or failure left you doubting everything, and you found yourself asking:

Why is this happening to me?

Why do I have to struggle while others seem to have it easy?

Why can't things just go right for once?

Hardship can also be relentless. In 2012, a motorcycle accident tragically took Fayçal's beloved younger brother's life. After months of handling the complexity of his brother's estate while supporting his parents through their grief, Fayçal poured his energy into cofounding an audacious consumer electronics start-up that caught the attention of Apple and Porsche. But then his most trusted business partners, people he thought of as older brothers from his childhood, left him in a devastating position. The fallout forced him to start over from scratch. As Fayçal remembers, "The betrayal was unbelievably painful, and restarting with nothing was one of the greatest battles of my life—all while still coping with my unreliable health. The stress led to flare-ups. It would've been so easy to slip into being the victim and ask, 'Why me? Why does misfortune keep following me? Why can't I get a break?'"

In his anger, grief, and chronic ill health, Fayçal could have believed the lie that his actions didn't matter. He could have listened to his self-doubt shout that he didn't have the power to change his circumstances. He could have let these inexplicable moments shatter his resolve, harden his heart, or make him bitter. But he chose a *better* path and found strength in the very heart of chaos.

As Fayçal said, playing the victim "wouldn't have changed a thing about my situation. I had to remind myself that the counter to the 'Why me' question is 'Why not me?'"

Becoming *Better*

Psychologists call Fayçal's approach to the life-defining challenges *post-traumatic growth*. These are positive psychological changes that can arise after experiencing trauma. Studies show that about 53 percent of people who experience trauma will find ways to grow stronger because of it.[1]

On a practical level, these choices are about how we deal with times "when hard chooses us." But on a deeper level, they are about *who* we become in the process.

The story consultant Robert McKee writes in his book *Story,* "True Character is revealed in the choices a human being makes under pressure—the greater the pressure, the deeper the revelation, the truer the choice to the character's essential nature."[2]

Trees in the forest are living proof that strength comes from struggle. Young saplings reach for the sun, converting its light into energy to grow. But when sunlight is constant and unfiltered, when growth comes too easily, the trees grow fast—but they're weak. Their wood is soft, shallow, and vulnerable to disease, mold, and decay. Forester Peter Wohlleben, in *The Secret Wisdom of Nature,* explains that trees growing in the shadows—fighting for every sliver of sunlight—develop denser, stronger wood. Why? Because resistance builds resilience. "Developing mighty trunks," he writes, "takes a great deal of energy." [3] It requires struggle.

This applies to us too. When everything comes easily, you don't develop the resilience to withstand real challenges. Struggle, discomfort, and effort shape you. They make you capable. They give you the mental muscle to take ownership of how you approach life. Just like trees that fight for light grow stronger, you grow stronger by facing resistance and persevering through setbacks.

Fayçal's story is a testament to this. When it came to his health, he understood his limitations, but he refused to let them hold him back from a fulfilling life and making a difference. Instead, he asked himself a simple but powerful question: "How can I make the best with what I have?"

When we face constraints head-on, something surprising happens: We become *more creative*. A 2019 review of 145 studies on the role of limitations on innovation found that limited options force us to think harder, adapt faster, and innovate smarter.[4] We actually come up with *more* and *better* solutions.

Constraints don't have to limit us; they can push us to be our most resourceful selves. This mindset is what's driven Fayçal over the last three decades, despite, as he puts it, "feeling shitty most of the time." Instead, he has always zeroed in on what he can control—with an unshakable belief that there's always a way. He constantly asks himself, "What steps *could* I take?" followed by, "What steps *will* I take?"

When he became the victim of fraud and lost everything, instead of succumbing to defeat, he returned to Australia and joined forces with a business owner in a completely new field for him—the fast-growing solar industry. Within six months, he had helped that business grow ninefold. In 2017, just months later, Fayçal and I met. Together, we've built not only a life but also businesses that are making an impact on people across the globe.

PRACTICE: Make "I Could" and "I Will" Lists

When you catch yourself focusing on how wronged you feel or how unfair a situation seems, use the occasion to shift gears. Replace *Why me?* with *What now?* to change what happens next. Even if you feel drained or lack clarity, interrupt any possibility of blame and frustration by taking small steps toward Autonomy.

Step 1: Start with an "I could" list

Brainstorm every possible step you *could* take, no matter how minor it seems.

Step 2: Then, create your "I will" list:

Scan your list, circle the actions you have control over, and commit to them.

For example, if you unexpectedly lose your job, your "I could" list might include reaching out to a mentor, setting up informal interviews, updating your LinkedIn profile, and looking into events you could attend. From this, your "I will" actions might be sending an email to a mentor today and updating your LinkedIn profile by the end of the week. This is how you reclaim control, by creating small, repeatable moments of Autonomy. Even in tough times, you can choose to move forward. And over time, this can become a habit—because each small action reinforces the belief that you *can* influence what happens next.

The Story You Tell Yourself About Your "Hard"

The hard things that happen to us, the mistakes we make, and the tragedies around us are real and sometimes devastating. But as we've seen, the story we tell ourselves to make sense of these hard experiences can deepen our pain, fears, and doubts. Or they can help us move forward.

How you explain the hard things matters more than you might think. It reflects what's called your "explanatory style,"[5] or the story you tell yourself about what just happened and why.[6] There are two types. A *pessimistic* style is where you explain setbacks as personal and pervasive. You don't just experience hard things; they become part of your identity. An *optimistic* style, on the other hand, is where you explain setbacks as temporary and specific. It creates separation: *That moment was hard, but it doesn't define me.* And that separation helps you face reality without letting it narrate your future.[7] Just ask Peter Best.

At nineteen, he was the drummer for a scrappy little band with big dreams. For two years, he rehearsed, performed, and planned a future alongside them. Then, just as the band was taking off, they replaced him—without warning, without explanation. Just . . . gone.

The guy who took his spot? A young drummer named Ringo. That band was The Beatles.

What followed was devastation. Peter spiraled into anger, depression, even a suicide attempt. One day, he was living the dream. The next, he was delivering bread while his former bandmates were becoming global icons.

But Peter Best didn't let that moment define him forever. Today, he tells a different story. "I'm happy. . . . I have no complaints. I've enjoyed life. Wouldn't change anything," he reflects.[8] He eventually left the performing world, worked in civil service, and made peace with his almost-fame and the stories he now tells. He's been happily married for more than fifty years, has two daughters and grandchildren he adores, and lives a life that feels full.

He knows his life would have been different if he'd stayed in the band. But if you carry around resentment, he says, "you're going to end up a bitter and twisted old git." So instead, he chose meaning: "All the things that have happened to me, good and bad, happy and sad, have made me what I am today."

Whether we realize it or not, we're constantly crafting the stories that define our lives—who we are, where we've been, and what it all means. This process is what psychology professor Dan McAdams calls *narrative identity,* a concept he's spent more than forty years studying.

When Autonomy is low, those stories slip into what McAdams describes as a *contamination arc:* "I had it good . . . and then everything fell apart." It's a powerless plotline, where suffering becomes proof that nothing ever works out.

But when you cultivate Autonomy, your personal stories shift. You start telling *redemption arcs*—narratives where struggles have meaning and give way to strength.

A 2024 review of sixteen studies found that the transformative power of our stories doesn't come from changing the facts.[9] What's done is done and you can't rewrite history. The power comes from changing the *meaning* you assign to those facts. You decide how the past shapes your future.

Fayçal's Algerian parents have a story that could easily have become one of resentment, disappointment, and loss. It would've been understandable if they had framed their journey as one of compromised opportunities—his father, a pioneering oil and gas engineer, and his mother, a highly qualified French teacher in the UAE, were forced to leave their careers behind when they immigrated to Australia in search of a better life for their kids. Their qualifications were only partly recognized. And on top of that there was another major hurdle: the language barrier. Instead of working in their chosen professional fields, they poured everything into running a small local business—seven days a week, with no prior experience, in a community that wasn't exactly welcoming to migrants.

And yet, they didn't tell a story of loss. They rewrote it as a story of purpose. His father, who had once applied himself relentlessly to his studies in order to escape poverty and the challenges of his upbringing, carried the same mindset into this new chapter: "When you make a decision, you go for it. No looking back."

Gradually, they invited people from all backgrounds into their lives. They made friends. Even when they faced racism—direct, unfiltered hostility—they met it with strength and compassion.

They rewrote their narrative in real time, from one of loss to one of love, resilience, and new beginnings. That became the foundation Fayçal carried with him as he faced his own life-altering challenges.

The highest performers we've worked with understand the power this gives them. They don't deny their pain—they've learned to alchemize it. They've turned struggle into strength, using it to shape themselves into better, wiser versions of who they are. As so aptly articulated in the Bahá'í writings: "The more you plough and dig the ground the more fertile it becomes. The more you cut the branches of a tree the higher and stronger it grows. The more you put the gold in the fire the purer it becomes. The more you sharpen the steel by grinding the better it cuts."[10]

Narrative identity reminds you that no matter where you start or how powerless you feel, you can always reshape your story.[11]

Every obstacle you face is one of two things: a reason to grow, or a reason to give up.

The choice is always yours.

PRACTICE: Rewrite Your Story

This practice is grounded in proven principles to help you gain clarity, externalize the weight of your challenges, and reshape your story into one of strength and growth.[12] When you engage with your struggles in this way, you can regain a sense of control over your life. And that control empowers you to shape a future fueled by intention and purpose.

Step 1: Share Your Current Story

Think back to a painful yet defining moment in your life—something that shaped your sense of identity. It might have been something big, like losing someone you loved, facing a natural disaster, staying stuck in a job you hated, being betrayed, making a costly mistake, or falling for the wrong person. Or it could have been something smaller but still powerful—like choosing a career to please your parents, saying no to an opportunity, or staying in a toxic workplace. Set a timer for fifteen minutes and write it all down as you'd describe it to someone else. If writing feels too formal, record it as a voice note.

Remember Bruno, the CEO you met in chapter 14 who was stuck in a blame-and-complain spiral? When Bruno was finally ready, we worked through this exercise during one of our sessions. He chose not one moment but a thread from his childhood. He shared:

> Growing up, nobody ever believed in me. My sister was the "golden child," and I was always ignored. My parents never supported me, and because I always felt different, I felt like I had to prove myself all the time. Life felt harder for me, and still feels that way. That's why I'm so driven now—I feel this mad rush to make something of myself.

If the experience of revisiting your defining moment still feels raw or overwhelming, give yourself a little extra care before moving on to step 2. Try this protocol: For four days in a row, set aside fifteen minutes per day to write about the same experience. Let it all out, every thought, feeling, and raw reaction, in as much detail as possible. Don't worry about grammar or structure, and you can even write the exact same thing each day. More than two hundred peer-reviewed studies confirm that this kind of expressive writing can help you process the experience and ease emotional stress.[13] It might feel intense, and that's okay. After the four days, give yourself at least a month to let things settle before moving to step 2.

Step 2: Externalize Your Story

Your story doesn't define you—it's something that *happened* to you. You can disentangle yourself, and this separation allows you to tweak or edit that story. You're not denying reality, simply deciding what meaning you want to pull from the experience. Ask yourself:

- *What's the story I'd rather be telling?*
- *How would I prefer to see myself handling that situation?*

Bruno's breakthrough came in his next session. Reflecting on the story he would prefer to be telling himself, and how he prefers to see himself coping with the situation, Bruno shared:

> I've always felt like an outsider, but that's not who I am. Those early experiences aren't my identity—they're just things that I lived through. I'm not that same kid anymore. The story I want to tell is that I faced tough things, but I grew from them. I made them count. Those challenges had a purpose, and I'm stronger because of them. They taught me to rely on myself.

Step 3: Re-Author Your Story

Reflection 1: Who were you before?

Look back on who you were *before* that defining moment.

- What were your values, beliefs, and identity at that time?
- How did you see yourself and your place in the world?

For Bruno, this was raw:

> I didn't have a place in the world. I had to fight for everything. It was really tough. I had no power as a kid, so I think

> that my tendency to complain about things was a coping strategy I learned—so much was out of my hands that complaining helped me feel a little more in control. It gave me a sense of release.

Reflection 2: Who are you now?

Think about who you became *after* that experience—who you are now.

- How did this experience influence your identity and who you are today?
- What inner resources did it help you develop?
- What important lessons did you learn, and how have these impacted your approach to life and future challenges?
- How do you see yourself now?

Use language that reflects insight, like "This experience taught me that," "I've learned," "The reason that," "It struck me that," "I now realize."

Bruno reframed his experience like this:

> This experience taught me I'm tougher than I give myself credit for. Growing up, I felt everything was harder for me, but those struggles forced me to be resilient and resourceful. I've learned to face adversity and come out stronger.
>
> I've realized that those hard moments were proving that I could take a hit and still move forward, not that I was powerless. Yeah, life can be brutal, but I've shown myself I can handle it. I don't see myself as the underdog anymore. I trust myself now. I know I can figure it out, no matter what.

Bruno didn't need the world to change—he needed to change the way he saw himself in it. And this process of externalizing and rewriting his narrative helped him do just that. With this shift, Bruno recognized he didn't have to carry it all alone. He hired a business coach to keep him in check, restructured his executive team, and focused on what he could influence.

• • • •

You can't rewrite history, but you can rewrite its meaning. The stories you tell about yourself either trap you in the same old patterns or give you the freedom to move forward. You need to let go of the outdated identity that no longer serves you and embrace the stronger, wiser version that made it through. This is how you reclaim your Autonomy and build Big Trust.

Life will throw challenges your way. The question is: Will you let them define you, or will you define them?

CHAPTER 17

The Gift of Hope

Autonomy in the Here and Now

In a world where so much is out of our hands, reclaiming Autonomy begins with accepting what you can't control while remembering that you're never powerless. On the one hand, you can focus on what you can influence, build your capacity to sit with discomfort, and rewrite the stories that keep you tethered to your suffering. But on the other hand, there's also a deeper path on the way to Big Trust: *hope*. It's an inner conviction and sense of possibility that things can be different.

When I look back through the research surveys, client notes, and transcripts of interviews, there's something that stands out. The people who kept moving forward, despite the setbacks and "hard things," weren't always the smartest, most talented, or most skilled. The *one thing* that seemed to truly set them apart was something else: They believed there was a way through. They had something deeper than confidence: They had hope.

I saw it again and again in their words:

"I don't know exactly how, but I know I'll find a way through this."

"There's always another door to knock on."

"I just need to figure out the next step, and then the one after that."

The late professor of clinical psychology at the University of Kansas Charles R. Snyder spent his career studying hope. He came to define it as a life-sustaining cognitive state—a way of thinking. He used this beautiful analogy to explain it: "A rainbow is a prism that sends shards of multicolored light in various directions. It lifts our spirits and makes us think of what is possible. Hope is the same—a personal rainbow of the mind."[1]

Hope, in other words, is knowing that the future you want is possible, even if you don't yet know how to get there. It's the fuel that keeps you moving forward, the light that helps you navigate darkness.

Honoring Hope

This is the essence of hope:
Direct your focus to what you can control and take intentional steps forward, driven by a strong belief in your ability to shape a better future.

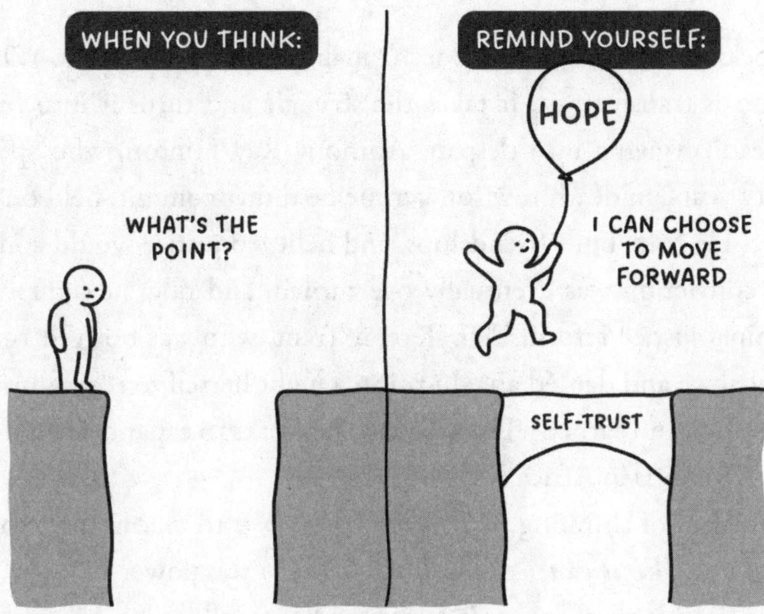

When life knocks you down—when rejection stings, illness looms, or loss pulls the air from your lungs—hope is what stops you from drowning in doubt. It helps you destroy the illusion that you can control absolutely everything and reminds you that you are *not* the most powerful force in the universe. That even in the worst moments, the story isn't over.

For many people, faith can help them experience hope or a mission that gives a sense of meaning to their struggles. Hope becomes a story, the arc of a rainbow, that helps us reconcile the past, take ownership of the present, and find the strength to move toward a better future.

Hope Is Belief in Learning over Lamenting

Hope doesn't erase pain. It doesn't make hardship disappear. What it *does* is transform it. It takes the struggle and turns it into fuel. Instead of giving into despair, Anthony Ray Hinton, who spent thirty years on death row for a crime he didn't commit, held on to hope. He read, built friendships, and believed justice would come. His conviction was eventually overturned, and now he fights for criminal justice reform.[2] Dr. Tererai Trent, who was born in rural Zimbabwe and denied an education, taught herself to read, moved to the US, and earned a PhD. Today, she works to expand education access for girls in Africa.[3]

Instead of thinking, *Why did this happen* to *me?* hope helped them ask, *What can I* do *with this?* That's the power of hope. It turns pain into purpose. It turns mishaps, failure, and hardship into wisdom.

As researcher and author Brené Brown is quoted saying, "One day you will tell your story of how you've overcome what you're going through now, and it will become part of someone else's survival guide."[4] I know I have those hard moments in my life that I'm still not ready to talk about, but they became bearable when I realized that I could write and speak about them in the future to help others.

I know that day will come. And I know the same is true for you.

Hope Is Belief in Your Present and Future

Hope doesn't just reshape the stories of what's happened to you in the past, it also anchors you in the now. Hope strips away your self-doubt by concentrating your mind on actions, choices, and solutions instead of ruminating on what's out of your hands. Hope also means that you don't live in *regret*—you live in *reality*.

If a path still matters to you, choose it now. If it's no longer what you want, let it go. Either way, stop standing at the crossroads. Make a choice. Move forward.

The point isn't the size of the action, or being reckless or motivated by pipe dreams. It's that you *take* an action. Let hope give you a belief in what's possible, and a belief that you can take ownership of your life.

Trust me—the return on trusting yourself and taking purposeful action compounds quicker than you would think. When you've made peace with your past and taken ownership of your present, something powerful happens: The future stops feeling like something that's happening to you and starts feeling like something you're building. You begin to trust that your choices matter. That change is possible.

Taking action on your own behalf requires Autonomy. And Autonomy depends on hope. When you choose to act, no setback is final, no suffering is wasted, and no future is predetermined.

What if hope is simpler than we think? Here's what it might look like:

- **For the past:** Remind yourself that nothing is wasted, not even the moments you wish had gone differently. Your experiences have shaped you, strengthened you, and prepared you for more than you realize. Let yourself believe that good can follow hard. You've made it through every single one of your hardest moments.
- **For the present:** Look for one small thing within your influence today. It could be a conversation, a choice, a quiet act of courage.

In every step forward, especially when the path is unclear, you're making a declaration: *I'm still here. I'm still building.* You don't have to see the whole sky to know the rainbow is coming. Just keep walking toward the light you *can* see.

- **For the future:** Hold space for the possibility that things can get better. Not because it's guaranteed, but because your next move still matters, and you're the one who gets to make it.

Whatever your goals, whatever you want to be different or better in your life, however much you want to move past self-doubt or simply find more joy, you have to *embrace your Autonomy.*

Every time you *move into life* rather than letting life move you, you tap into Big Trust, where you let go of self-doubt and take ownership of the life you're creating. And hope is the catalyst that helps you do it.

TAKE A MOMENT:

Imagine what you could achieve if you fully embraced Autonomy, channeled hope, and made a choice to move forward today, this week, and beyond.

The Fourth Attribute

ADAPTABILITY

CHAPTER 18

Can I Manage My Emotions?

The Core Question of Adaptability

Mira, a thirty-eight-year-old communications freelancer, was at a client dinner—the kind where you're technically "off the clock" but still quietly pitching yourself for future work. Midway through the meal, one of the senior executives made a subtle dig about freelancers being "lucky to get consistent work these days." He gestured toward her as he said it.

"I could feel my stomach drop," she told me later. "My face got hot, my heart was racing, and I had at least three sarcastic replies ready to go."

But instead of snapping back or shutting down, Mira took a breath, reminded herself of the bigger picture, and simply smiled. "You're right—it's a competitive space. Luckily, I love a challenge."

Afterward, she reflected: "I felt it all—rage, annoyance, defensiveness, embarrassment—but I didn't let it hijack the conversation."

Mira finished the dinner calmly, stayed engaged, and even secured a follow-up meeting with another executive at the table. She didn't ignore her emotions. She noticed them but didn't let them take over.

That's what Adaptability does for you. Mira was able to set her self-doubts aside and stay grounded—even in the face of dismissiveness. It helped her respond with composure instead of reactivity. And it paid off.

Remember at the beginning of this book how we explored the difference between "good" self-doubt—the kind that sharpens your focus, encourages exploration, and pushes you to grow—and "bad" self-doubt, the kind that paralyzes you, makes you feel inadequate, and keeps you stuck?

Emotion is the one factor that sets these two experiences apart.

Think about it. Every moment of self-doubt you've ever faced has been charged with some sort of *feeling*. It might be the dread curling in your stomach before a big interview, the spike in your heart rate when you're about to share something important, or the heaviness that sits on your chest when you scroll through LinkedIn and suddenly feel behind, lesser, not enough.

It's not the doubt itself that defines your experience—it's how you *feel* about the doubt. And more importantly, how well you can *stay grounded* in the middle of that emotion. And that's a choice.

Why Groundedness Matters

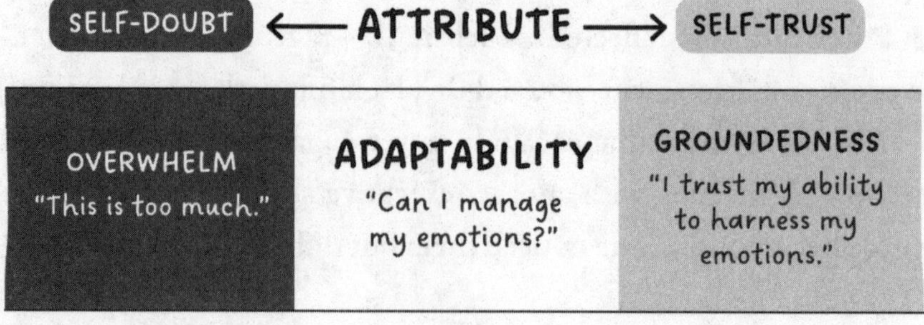

The search for groundedness is the defining challenge of Adaptability, the fourth Attribute. In the context of Big Trust and the Doubt Profile, Adaptability reflects how well you're able to regulate your emotions and maintain a steady internal state, especially during moments and experiences that challenge us.

When Adaptability is weak, self-doubt feels like a tidal wave—pulling you under, making every setback feel personal and every challenge feel overwhelming. Instead of moving *through* your emotions, you get stuck in them. When you struggle with Adaptability you don't question your worth (that's Acceptance), your abilities (that's Agency), or your influence over the situation (that's Autonomy). Low Adaptability says, "I can't handle how this feels without falling apart."

And when emotions take over, you get reactive. You snap, withdraw, overthink, or spiral. Everything becomes urgent. Nothing feels manageable.

If that sounds familiar, you're not alone. Our research shows that 52 percent of people fall into the Red Alert or Hindrance zones for the Attribute of Adaptability—meaning that more than half are struggling to stay grounded when things go wrong. Another 17 percent land in the So-So zone, managing emotions sometimes, but not consistently.[1]

In the first section of the book, "The Self-Doubting Mind," we covered how emotions have a strong evolutionary purpose and how our stories and individual Doubt Profiles help or hinder our ability to harness them.

Emotions are essential. They keep us alive. They flag potential danger, prime us to act, and help us navigate the complexities of being human. But in our lives today, those same emotions often misfire, leading to confabulated worries and self-doubt that can snowball into a full-blown existential crisis. As psychologist Daniel

Goleman writes in *Working with Emotional Intelligence,* "Out-of-control emotions can make smart people stupid."[2]

The emotions associated with self-doubt can drag in a whole crowd of feelings: shame, anxiety, stress, fear, resentment, disappointment, and more, depending on which Attribute you're struggling with most in the moment. These emotions are all tied to how you *perceive* threats in your life. And these emotional states not only reflect your self-doubts but also fuel them. Each one feeds the next, creating a vicious doubt-emotion cycle that keeps you stuck.

THE VICIOUS DOUBT-EMOTION CYCLE

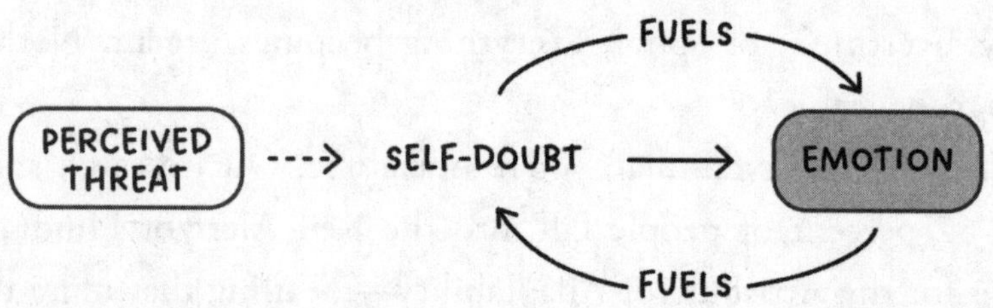

Think about how many times you've been caught in that loop—where your doubts trigger intense emotions, and those emotions only make your doubts feel more real. Like when you:

- **Let one bad moment ruin an entire day,** replaying it in your head and feeling frustrated or humiliated until it felt bigger than it really was.
- **Struggled to separate emotions from reality,** assuming that because you *felt* incapable, you *were* incapable.

- **Avoided challenges that made you uncomfortable,** not because you lacked skill, but because you dreaded the emotional toll of discomfort, failure, or judgment.
- **Felt consumed by stress or anxiety,** because you either didn't know how to process it or believed that worrying nonstop somehow meant you cared enough.

The real challenge of low Adaptability isn't so much the moment itself. It's whether you trust yourself to handle the emotions that come with it.

The Drive for Emotional Equilibrium

As humans, we're wired with a deep drive for emotional equilibrium. Your brain and body are constantly working to maintain balance through a process called *emotional homeostasis*. It's that steady internal state where we feel balanced and in control. We crave it. And when it's disrupted, our system scrambles to restore it, often through unhelpful patterns like avoidance, lashing out, or shutting down.

How we chase that equilibrium is shaped by years, even decades, of learned responses. If you grew up brushing off disappointment, stuffing down frustration, or walking on eggshells to keep the peace, those habits stick. They become your go-to responses when life gets uncomfortable. If you grew up in a household where emotional outbursts were the norm, you might have learned that the only way to feel seen or heard was to explode. Over time, that kind of reaction can feel like the fastest route to restoring a sense of control or dignity. And the more you react that way, the more habitual it becomes.

Dealing with emotions can feel so personal, and so overwhelming, because it often feels like they are happening *to* us, not *with* us. They can feel out of our control and feed our self-doubt. In my research, I found that difficulty managing emotions usually goes hand in hand with feeling disempowered.[3] The more out of control your emotions feel, the harder it is to believe you can choose how to respond, and this chips away at your Autonomy.

Some people are biologically more emotionally steady than others—owing to differences in temperament and the sensitivity of our nervous systems. But for all of us, emotional adaptability is largely learned. We absorb it from the way emotions are handled in our homes, the reactions modeled around us, and the unspoken rules we grew up with.[4]

Those rules are often shaped by gender and culture. When you were growing up, maybe you were taught that boys shouldn't cry or that girls shouldn't be too assertive, angry, or "difficult." Perhaps you received messages to "be nice," "be strong," or "not make a scene." Whether it was said outright or indirectly, the result was the same: You learned that some emotions are acceptable and others are not. When cultural expectations come into the mix, the emotional rulebook gets even more complicated. In some cultures, emotions are met with patience and openness, which builds resilience. In others, expressing emotion is seen as indulgent, weak, or even shameful. Maybe you were told to "stop being dramatic" or "toughen up," so you did. And somewhere along the way you started to believe that *feeling too much* meant something was wrong with you.

As an adult, those early lessons show up at work, in conflict, under stress, and when facing something hard.[5] In those moments, you're not just reacting to *now*; you're reacting with a lifetime of conditioned responses, powered by a nervous system that's simply trying to protect you the only way it knows how.

And as if that isn't challenging enough, research is beginning to show that the ways our brains are wired to handle emotions intensify our self-doubt and interfere with our ability to calm ourselves.

1. Meta-Emotion Mayhem. Finding emotional equilibrium is made so much harder because our emotions actually judge each other.[6]

You're anxious . . . then anxious about being anxious.

You feel self-doubt . . . then get frustrated that you're doubting yourself.

You're stressed . . . then stressed that you're so stressed.

And let's not forget the classic: You worry . . . then worry about how much you worry.

Here's what it might look like. You're prepping for a big pitch, and a flicker of self-doubt creeps in: *What if I mess this up?* Totally natural. But then, *I shouldn't be this nervous—I'm a senior leader. What the heck is wrong with me?* Now you're not just tackling the doubt, you're drowning in shame and self-judgment. This is the self-perpetuating cycle of meta-emotions. The more you fight how you feel, the harder it gets.

2. "Big Feelings." Since some people are wired with a more sensitive nervous system, you may feel like your emotions are at full volume while everyone else is humming along. In fact, you might be a "Highly Sensitive Person" (HSP), a term coined by leading psychologists and researchers Elaine and Arthur Aron. The Arons estimate that 20 to 30 percent of people fall into this category.[7]

If you are somewhere on the HSP spectrum, you process emotions more deeply, notice subtleties others miss, and absorb other peoples' energy like a sponge.[8] Every negative emotion is more intense.[9] You *feel* more deeply than others. You're naturally more reactive and more easily stressed, so things can feel overwhelming much more quickly. Your brain is wired for hyper-awareness, what scientists call *sensory processing sensitivity*. This means you may over-analyze, over-worry, and yes . . . over-doubt.[10]

If this sounds like you, there's a good chance a parent or grandparent experienced the same sensitivity (in my case, thanks, Mom).[11] If you grew up hearing, "You're too sensitive" or "Stop overthinking," you probably started to second-guess every emotion you felt. The result is self-doubt. A lot of it. But still, you can learn to use awareness of your sensitivity to create distance between what you feel and how you respond.

3. The Fix-It Trap. Your brain is wired to help you avoid threats and seek rewards. That means it is constantly scanning your environment (and your inner world) for signs of danger, and this includes emotional discomfort. When it detects something unpleasant—like anxiety, fear, shame, or self-doubt—it tries to "solve" it quickly to return you to a sense of safety. But this creates a reflex to *fix* emotions rather than feel them. That's why we instinctively reach for quick relief: swipe away the discomfort, distract ourselves with dopamine hits, or shut it down with logic. We've been conditioned to treat discomfort like a glitch in the system.

However, when you suppress or resist uncomfortable thoughts and feelings, your Gatekeeper (your brain's filtering system) can keep flagging them as important and bringing them back into your awareness. Psychologists call it the *rebound effect*.[12] The more you try to avoid discomfort, the more intensely you might find yourself focusing on and experiencing it. Over time, this can wire your brain to see struggle as a signal that something's wrong with *you*. That if you're not happy all the time, you're failing. But trying to live that way is exhausting.

Yes, real dissatisfaction deserves your attention. But emotional discomfort isn't always a sign that something's broken. Sometimes, it's just a sign you're human.

Emotions don't need to be neat and tidy. You just need to catch them before they spiral into self-doubt—and harness them so they can guide you, rather than pulling you under.

The Promise of Adaptability

When you're emotionally grounded, you're able to calibrate or adapt your response and feel a sense of control over your feelings. You're not ruled by every emotion that comes up. You don't snap in anger, spiral in anxiety, or shut down in sadness. Instead, you recognize your emotional triggers. You pause. You assess. You choose how to respond—on purpose.

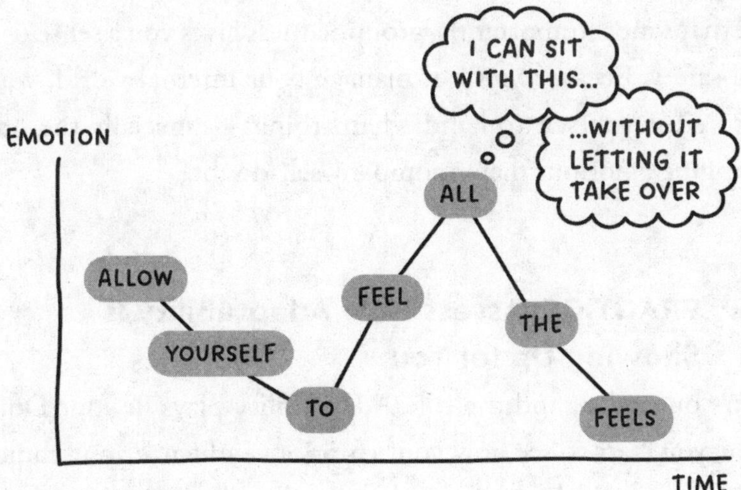

When your Adaptability is strong, you stay connected to yourself *through* emotion. You can *feel* discomfort and fear without losing your footing and feeling you're incapable. You can *feel* the sting of rejection without making it mean you're unworthy. You experience emotions fully—without being overwhelmed by them.

This allows you to manage stress with composure and confidence, which means you perform better at work.[13] Adaptability is closely linked to emotional intelligence,[14] helping you express what you're feeling in ways that are clearer, more constructive, and easier for others to connect with. That kind of emotional clarity strengthens your relationships, whether with a partner, friend, or colleague. And when life gets chaotic, Adaptability is what helps you stay calm and clear, which is a powerful driver of long-term happiness.[15]

As Mira demonstrated, emotional groundedness can be learned and practiced—even when your nervous system is on higher alert. You can experience your emotions, tap into your Big Trust, and express yourself. And when you mess up, blurt out something, or are suddenly hit with a feeling of overwhelm, you can quickly regroup and keep moving forward without getting trapped in a spiral of self-doubt

Perhaps most importantly, groundedness gives you a sense of internal safety. Because you can manage your internal world, you're able to trust yourself to handle hard things—especially the often challenging emotions that accompany self-doubt.

PRACTICE: Assess How Adaptability Is Showing Up for You

When you understand the role Adaptability plays in your Doubt Profile, you start to see how your emotions influence your choices, reactions, and self-doubt. From there, you can begin building habits that help you stay grounded, regulate what you feel, and respond with clarity—even when self-doubt hits.

Look back at your Adaptability score from the Doubt Profile diagnostic. What was it?

Adaptability Score: __________ Zone: __________________

(For Zone, write in "Red Alert," "Hindrance," "So-So," "Hidden Strength," or "Superpower.")

Take some time to reflect on the questions below. You might like to journal your responses:

What are your vulnerabilities around the search for Adaptability?

- **Think back to childhood—how were emotions handled at home?** Were you encouraged to express them, or taught to suppress them? How did those early messages shape the way you manage difficult feelings today?
- **How do your emotions fuel your self-doubt?** Do you spiral when something goes wrong and feel like you can't do it? Struggle to separate your feelings from your actions? Shut down when emotions feel too intense?
- **When was the last time you were overwhelmed but found a way through?** What allowed you to regulate your emotions instead of letting them dictate your actions?
- **How might your life look and feel different if you had more trust in your ability to ride out tough emotions?** What would it unlock for you?

When your Adaptability feels shaky, what strength can you lean on?

- If you feel stuck in overwhelming emotions, you can draw on **Acceptance**. Emotional struggles are a natural part of being human, not proof that something is wrong with you. What would it look like to offer yourself the same kindness you'd give a friend in a difficult moment?

- If you're caught in reactive patterns, lean into **Agency**. What's one emotional regulation strategy you've used before that helped you regain control—breathing exercises, movement, journaling, or something else? How can you use that tool today to remind yourself that you already have the capacity to manage what you're feeling—and trust the proof you've done it before?
- If emotions are clouding your judgment, turn to **Autonomy**. Remind yourself that you don't have to be at the mercy of your emotions. Focus on what you can control in the moment and commit to one choice that helps you move forward, even if it's small.

• • • •

In chapters 19 and 20, we'll dive into ways you can harness your emotions and turn them into fuel for resilience rather than resistance. You'll learn practical strategies to help manage the emotional waves that throw you off balance and ways you can become more emotionally grounded. Then, in chapter 21, we'll uncover how to tap into a powerful emotional reserve that can't be manufactured—one that helps you build stronger Adaptability.

When you stop fighting your emotions and start working with them, your self-judgment eases. Your self-confidence feels more grounded. Your performance feels more fluid. And the weight of self-doubt feels a whole lot lighter.

CHAPTER 19

Can I Believe What I Feel?

Hear the Message of Your Emotions

When I was eight, my favorite song was Jewel's "You Were Meant for Me." I played it on a loop on my pink Discman, blissfully unaware that Jewel Kilcher, at the same age, had been abandoned by her mother.

By age fifteen, she'd left home to escape her abusive father. She lived out of a car, scraped by with odd jobs in San Diego, and battled anxiety and constant stress. But she developed a mindset that helped her survive.

She learned to see her emotions as *messengers*.

"If I eat bad fish and I get food poisoning and I throw up, something's not wrong with me; something's *right* with me. My body is working properly," she once said. "Anxiety, fear, stress, doubt are the same way. They're a neon sign trying to tell me that I'm consuming something—a behavior, a thought, or action—that my body doesn't agree with."[1]

Even as a teenager, Jewel understood something most adults struggle with: Emotions aren't defects. They're messages. Your emotions, especially the messy, inconvenient ones, aren't random. They're "neon signs" that something important needs your attention.

Often, these intense feelings point out that something essential to who we are—our identity, values, or needs—is being challenged or ignored. A pang of guilt might mean you crossed a line. A flare of anger might mean someone else did. A knot of unease might be your body warning, "This isn't right."

When you struggle with Adaptability, you misread the signal. It's not something you consciously do. Your system does it for you. Discomfort shows up—a tense conversation, a setback, a moment of uncertainty—and your automatic response hijacks the meaning. Instead of interpreting it as "This situation is hard," it becomes: "*I'm* the problem." Or worse, the signal gets ignored completely. Brushed aside. Suppressed. Dismissed. You jump directly to the self-doubt spiral and lose your grounding.

But there's another way. Gianpiero Petriglieri, an associate professor at INSEAD, puts it simply: "Emotions are data."[2] Not "good" data or "bad" data, just raw information that needs processing. Instead of reacting to every little thing or jumping to conclusions, the goal is to decode it. Whatever your emotional makeup or starting point today, you can develop the conscious habit of asking yourself, "What is this feeling trying to tell me?"

Think of emotions like the little notification pings on your phone. Each one—joy, jealousy, anxiety, sadness—is your brain sending you a message: *Hey, notice me!* Just like with our phones, you have a choice. Do you immediately swipe, tap, and get lost in the reaction? Or do you pause, take a breath, and get curious: *Interesting, what's this trying to tell me?*

To answer that question, you need to know and name what you feel.

Name Your Emotions

When I work with leadership teams, I often hear leaders say, "I'm stressed." It sounds so straightforward, and early in my coaching days, I'd take it at face value. I'd offer up time-blocking tools, mindfulness techniques, or productivity strategies. Sometimes they helped. But more often, they didn't.

Why? Because "I'm stressed" is almost always the tip of a much deeper iceberg. Beneath the stress are the deeper raw emotions that people haven't named—or don't feel safe enough to name. Things like:

> "I'm miserable and stuck in a career that's going nowhere. Every day, I wake up with this sinking feeling that I'm being judged, overlooked, or seen as a failure."

> "I don't think I'll ever fit in. I try so hard to be liked, to be competent, to be enough, but I always end up feeling like the outsider."

> "I don't even know what the point of my life is anymore. I'm constantly busy, but underneath the meetings and deadlines . . . I feel lost and almost panicky."

No amount of calendar optimization is going to solve existential dread. Susan David, a Harvard Medical School psychologist, refers to these catchall emotional labels (like, "I'm stressed") as overarching *umbrella terms* that miss the nuance of what we're really feeling.[3]

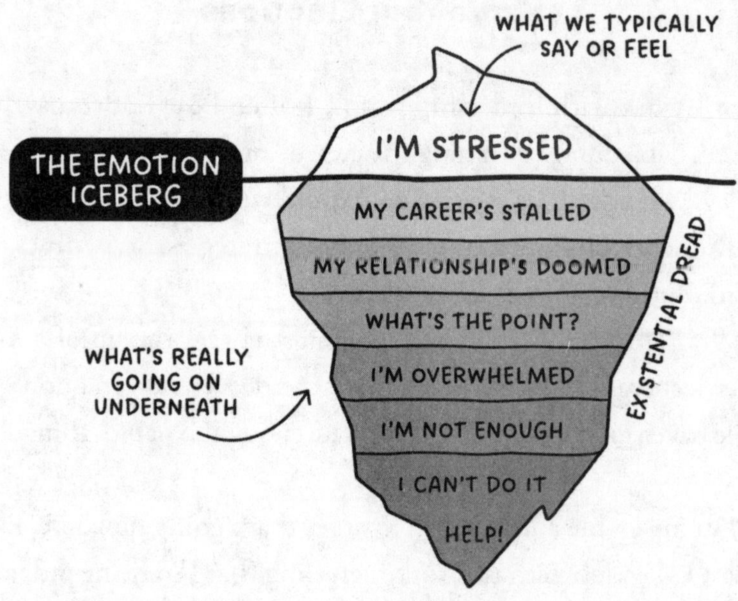

It reminds me of one of many classic dad jokes my father loved to tell my brother and me when we were kids:

A man walks into the doctor's office, visibly distressed. He sits down and explains his peculiar predicament.

Patient: "Doctor, every time I drink coffee, I get this sharp pain in my eye. I've tried changing brands, using different cups, even going decaf, but nothing helps."

Doctor: "Walk me through what you do when you prepare and drink your coffee."

Patient: "Well, I pour boiling water over the coffee

grounds in my favorite mug, add a dash of milk, give it a good stir, and then . . . I take a sip. That's when it hits—a sharp, intense pain in my eye."

Doctor: "Your prescription: Remove the spoon from the mug."

Cue my dad's hearty chuckle

Without clarity, we're like the guy stabbing himself in the eye with a spoon and blaming his coffee. We're walking around with discomfort, stress, or dread—and we misdiagnose the cause. We say it's "stress," "anxiety," or "burnout," when underneath, it's fear of failure. Rejection. Shame. Self-doubt.

So, how do we figure out what we really feel? We look somewhere else and start with our physical self, not our emotional one.

I learned that one of the most powerful questions I could ask a client, especially when they were overwhelmed, was: "Where do you feel this in and on your body?" Not "Why are you feeling this?"—but *where*. Because emotions show up in our bodies first, often before we can name them; they are physical *before* they are verbal.

"My chest feels tight, like it's caving in."

"I've got this heat in my throat that makes me want to scream."

"My stomach is in knots, like I'm bracing for a punch I can't see coming."

When we shift attention from the story in our mind to the physical sensations in our body, the emotion loses its grip. The anxiety eases. The overwhelm becomes less, well, overwhelming.

When you tune in to your body, you create a pause, and that pause opens up space to observe what you're feeling without judgment. It interrupts the loop of automatic thoughts that make an emotion feel all-consuming and prevents meta-emotion mayhem where your emotions start judging each other.

Here's why this happens: Your brain processes emotions through a sensory system called *interoception*[4]—which is your ability to sense internal signals like heartbeat, tension, or breath. When you bring attention to those signals, you start building what's known as *emotional granularity*[5]—the ability to name and distinguish what you feel.

So when you ask yourself, "Where do I feel this in and on my body?" you're becoming curious about your internal world. And in that curiosity, it becomes easier to manage whatever it is you're feeling.

As Dr. Daniel Siegel suggests, you have to "Name it to tame it."[6] Neuroimaging studies show that when you're experiencing an unpleasant emotion, simply naming what you feel reduces reactivity in the amygdala, dialing down the emotional intensity.[7] UCLA psychology professor Matthew Lieberman uses a memorable traffic analogy: "In the same way you hit the brake when you're driving when you see a yellow light, when you put feelings into words you seem to be hitting the brakes on your emotional responses."[8]

When we can't name or differentiate our emotions, we can't deal with them effectively—or ask for the right kind of help. The emotion will keep poking you in the metaphorical eye, and you keep wondering why life hurts so much.

Map Your Feelings

Years ago, I met Cait, a venture capitalist who had an intriguing daily habit of scheduling her emotional labeling. This sounds far more complicated than it actually was. During an afternoon catch-up, her phone buzzed with a notification. She explained that every day at 4 p.m., her phone prompted her with a reminder: "Check in with yourself." She'd pause for a moment to reflect on two questions: "What am I feeling right now?" followed by "Is this emotion serving me?"

For example, if she felt calm and attentive during a negotiation, great—she knew she was in the right state to make a smart decision. But if she felt impatient or distracted, she'd pause to reflect, "What can I do to refocus and give this the attention it deserves?" Sometimes that meant taking a short break or going for a walk. Other times, just a minute of deep breathing.

This tiny daily ritual kept her emotionally grounded and ensured she wasn't letting unchecked feelings drive big decisions.

Of course, sometimes it's hard to know exactly what you're feeling. Emotions can be slippery like that—especially when they're tangled up with stress, pressure, or a whirlwind of mental clutter. That's where a more refined self check-in can make all the difference.

Think of your emotions as a two-part story: There's what you *feel,* and then how you *interpret* it. According to neuroscientist Dr. Lisa Feldman Barrett's research, your brain is like a black box flight recorder. It's constantly collecting signals from your body—heart rate, breathing, muscle tension. Then, based on your past experiences, it tries to make sense of those signals in the moment.[9] That flip in your stomach? Your brain decides if it's fear, excitement . . . or just "bad fish."

And it does this in two acts:

Act One: *How intense is this feeling?* This is what scientists call *physiological arousal.* It's the raw energy you feel in and on your body. The rush of blood to your face, the racing heart, the knot in your stomach, the sweaty palms, or the warmth you feel that spreads throughout your chest. These sensations are powerful visceral messengers. They arrive *before* your brain has had time to explain what's going on.

Act Two: ***Is it pleasant or unpleasant?*** Once your brain has taken note of your arousal (or the intensity of the emotion), it still needs to make sense of what this means to you. This is *valence*. It relates to how it *feels* to you in the moment. Does it feel energizing or draining? Light or heavy? Context matters here. Where are you? Who's with you? Are there risks you're picking up on? A racing heart might feel exhilarating if your partner is on one knee proposing, but downright terrifying if you're called on to explain an issue to your boss.

When you break emotions down into these two simple dimensions—arousal (how intense it feels) and valence (how pleasant it feels)—it becomes a lot easier to make sense of what's going on inside.

This is where the Emotional State Matrix comes in. It's a simple way to map how you feel and understand how your emotional state might be fueling your self-doubt:

- **Top left—High Intensity, Unpleasant:** Think fear, frustration, and anxiety. You're buzzing, but not in a good way.
- **Top right—High Intensity, Pleasant:** This is your joy, excitement, and enthusiasm zone. Energy's high and it feels great.
- **Bottom left—Low Intensity, Unpleasant:** Emotions of sadness, loneliness, or insecurity. You're feeling low and heavy.
- **Bottom right—Low Intensity, Pleasant:** This is where you're feeling content, serene, and calm. You're steady and centered here.

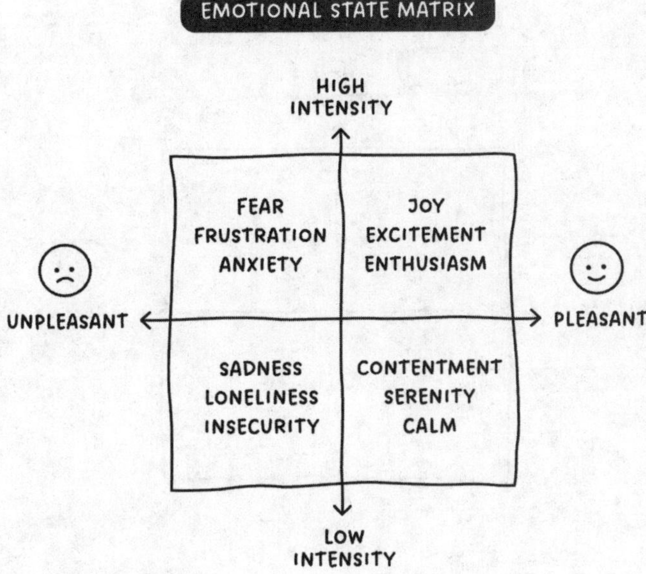

This idea of naming and taming emotions can be woven into your daily routine.

PRACTICE: The Self Check-In

Here's a simple two-step habit to boost your emotional Adaptability in real time. You'll name what you're feeling, locate it on the Emotional State Matrix, and decide what to do next—just like Cait did with her daily 4 p.m. prompt:

Step 1: Name It and Plot It

At any point in your day (set a reminder if it helps), pause and ask yourself: *What am I feeling right now? How intense is it? Is it pleasant or unpleasant?* Use the Emotional State Matrix to plot your emotion:

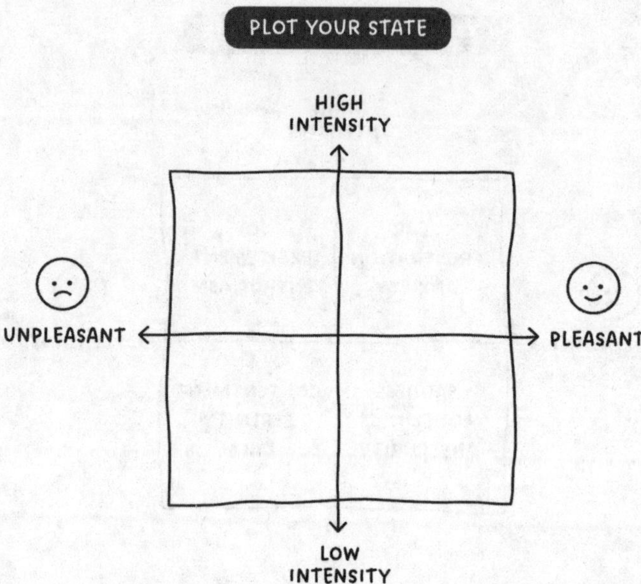

Step 2: Ask, "Is This Emotion Serving Me (or Others)?"

Not every emotion needs to serve *you*—sometimes it serves others by signaling care or concern. But it also shouldn't keep you stuck or spinning. Is your current state fueling clarity, energy, or motivation to move toward what matters? If not, explore ways to shift your state. Tap into your Autonomy and ask: *What small shift could help me feel better and move forward? Who or what could support that shift?* Then, take action.

And here's something important to remember: Not all "unpleasant" feelings are bad, and not all "pleasant" feelings are helpful. In fact, feeling *too* good can sometimes dull your thinking.[10] Research shows we're often at our most creative not when we're euphoric, but when we're a little uncomfortable.[11] So instead of judging what you feel, get curious. Emotions are information. Use them.

This quick check-in is a gentle way to make sure you're consciously shaping how you feel and responding to life's moments.

Separating *You* from Your Emotions

Earlier, I shared how leaders would often walk into a coaching session and announce, "I'm stressed." But there's more to unpack here than just the tip of the emotion iceberg. There's another trap we don't talk about enough: We don't just feel emotions—we *become* them.

When emotions run high, it's easy to say, "I'm anxious," or "I'm overwhelmed." It feels natural, even harmless. But when you say, "I am [emotion]," you're not just describing what you're feeling, you're defining yourself by it. Psychologists call this *cognitive fusion*.

It's the difference between saying, "I'm feeling stressed" and "I *am* stressed." One is temporary. The other becomes who you are. And over time, that language wires your brain to believe that emotions define you. That you are at the mercy of whatever feeling shows up.

You can go a step further to peel back your overwhelm and self-doubt. Once you notice and name your emotions, you can use the power of language to create space between *you* and your inner experience. Instead of becoming the emotion, you observe it, which practitioners of acceptance and commitment therapy (ACT) call *cognitive defusion*.

Rather than saying, "I'm anxious," you might say, "I'm *noticing* feelings of anxiety."

Instead of, "I'm stressed," you might try, "I'm *feeling* tension in my chest."

This subtle language shift helps you see the emotion as a temporary experience rather than something fused with your identity.

This technique also works with your thoughts. Instead of, "I'm a failure," you might say, "I'm noticing that I'm having the thought that I'm a failure." Yes, it feels a little clunky or robotic at first, but that's almost the point. This deliberate phrasing creates just enough distance between you and your emotions (or thoughts) to see them for what they are: passing experiences, not your whole identity.

Every time you do this you're reminding yourself, *This is something I feel. Not something I am.* Once you've created that space, you can see more clearly and choose your next move with intention.

Listening (Instead of Reacting) to Your Emotions

There's a persistent myth that emotions on the "unpleasant" side of the Emotional State Matrix are obstacles to success or happiness. That we should avoid, fix, or suppress them as quickly as possible. I touched on this briefly earlier. The truth is, these emotions often carry the exact insight we need to move forward. Even sadness can sharpen your focus and deepen your thinking.[12] Emotions like guilt, anger, and unease exist for a reason.

When we shut out or ignore these emotions, we dull the discomfort, but we also silence the signal telling us where to look and what to change. Worse, avoidance often leaves us feeling more stuck—disconnected from what's really going on and unable to course correct.[13]

Big Trust means having the courage to listen to what your emotions are trying to tell you, even when it's uncomfortable. It's trusting yourself to face the signal, not just soothe the symptom, because those signals are often at the root of our self-doubt. They flag when our actions and core values are out of sync. Once you start listening to the data your emotions provide, you can make choices that align with your true priorities. And alignment is a powerful buffer against self-doubt.

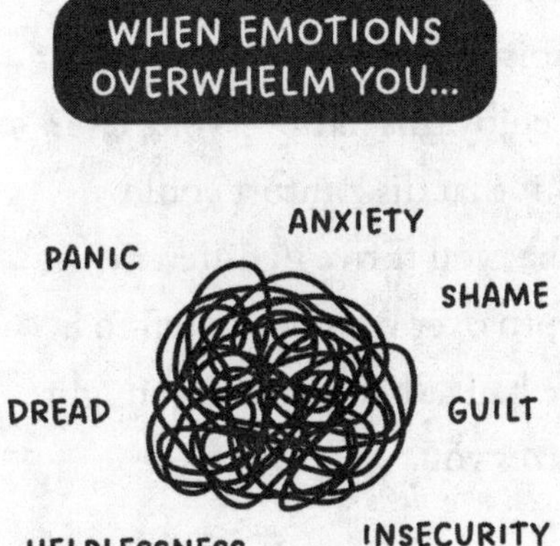

...UNPACK THE MESSAGES

- ✓ WHAT EMOTION IS THIS?
- ✓ WHAT'S THIS TELLING ME?
- ✓ WHAT NEED IS UNMET?
- ✓ WHAT CAN I DO RIGHT NOW?
- ✓ HOW CAN I REALIGN MY ACTIONS & VALUES IN THE FUTURE?

Here are a few examples of how you can shift from *reacting* to your emotions . . . to *listening* to them. When you give attention to the harder emotions (the ones on the left side of the Emotional State Matrix), it can ease your overwhelm and doubt, and guide you toward more meaningful and helpful responses:

- **If you're feeling guilt** about missing your kid's recital because of a work trip, it's easy to spiral into "I'm a terrible parent" and doubt your worth (low Acceptance). But pay attention to that guilt. It's a signal that your values are out of alignment and reminds you that family matters deeply, nudging you to realign your time with what's most important to you.
- **If you're feeling anger** about an unfair decision at work, that emotion might mask a deeper sense of powerlessness (low Autonomy). Instead of turning that frustration inward, ignoring it, or doubting your impact, recognize that it's

pointing to your sense of justice.[14] It's a cue that something needs to change, and you're the one to speak up.

- **If you're feeling uneasy** in a big, noisy group setting, you might start wondering, "Why can't I just enjoy this like everyone else? Why am I so socially awkward?" But that discomfort could be your nervous system signaling that you thrive in different environments—ones that value depth over volume. Listen to it. It's not necessarily a sign that you're bad at being social. It might be a cue to honor what truly energizes you.

At the end of the day, it's not the emotion itself that matters most—it's listening to it so you create the space and awareness to *choose* what to do with it.

PRACTICE: Uncover the Reason for the Emotion: What's the True Story?

When you're feeling overwhelmed by the emotions of self-doubt, there's huge value in stepping back to uncover the deeper reason for the emotion. Sometimes what you think you're reacting to isn't the full story. Understanding the deeper reason behind what you're feeling can be the key to regaining control and clarity because it allows you to *do* something with that information. This exercise will guide you through the process of uncovering those deeper reasons and connecting your emotions to your core values and sense of self.

Step 1: Pinpoint the Emotion

Start by naming exactly what you're feeling. Be specific. Are you feeling anger, sadness, anxiety, or something else? You might say, "I'm noticing that I'm feeling anxious" or "I'm carrying frustration."

Acknowledge these emotions without judgment. You can use the Emotional State Matrix from earlier if it helps.

Step 2: Explore the Trigger

Reflect on what you're responding to. What was the stimulus that sparked the emotion? Was it a snarky comment? A situation spiraling out of control? Something deeper?

- If you feel frustrated about constant interruptions at work, is it really about the interruptions, or is it about feeling disrespected or insecure about your ability to communicate clearly?
- If your team's lack of commitment makes you feel anxious, is it really about them, or does it tap into fears about your leadership or your business's future?

Step 3: Connect the Attribute

Emotions don't show up randomly. They're tied to something important to you. Figure out what part of your core self feels under attack:

- **Acceptance:** Do you feel invisible or unappreciated? Are you questioning your worth?
- **Agency:** Are you doubting your abilities or competence to handle something?
- **Autonomy:** Do you feel restricted, like someone else is calling the shots in your life or at work?
- **Adaptability:** Are you struggling to manage stress or emotions, making everything feel harder than it should be?

Naming the challenged Attribute helps you pinpoint the *why* behind your emotion—and that clarity is gold.

Step 4: Ask, "What's the Story I'm Telling Myself, and What's the *True* Story?"

If you're doubting your worthiness or capabilities, pause and ask if that story is really true. For example, if you're thinking, *I'm always interrupted because I'm not respected and bad at communicating,* invite your Fact Checker into your brain's boardroom (that's your internal voice of reason, the one trained to challenge confabulated stories with cold, hard truth). Ask: *Where's the evidence? What else could be true? What would a neutral observer see here?* Could it be that your workplace is a high-pressure environment and people just don't realize they're interrupting?

Challenge your first draft of the story. Look for alternative explanations. You'll be amazed at how often your initial take is completely incorrect.

Step 5: Do Something

Decide on a course of action that reaffirms your core values and strengthens the challenged Attribute. Ask yourself:

- How can I respond in a way that reinforces my Acceptance, Agency, Autonomy, or Adaptability?
- What's something I can do that aligns with my values and strengthens my self-trust?
- What support or resources do I need to help me move forward?

Then, take one meaningful step.

• • • •

Emotions aren't the enemy, or overwhelming by nature. They're messengers—signals pointing you toward something deeper: a value being neglected, a boundary being crossed, or a need going unmet. When you learn to pause, plot what you're feeling, and ask whether it's serving you, you stop being ruled by the emotions of self-doubt and start learning from them. You can then uncover the deeper meaning and take aligned action. This is how you build a habit of emotional adaptability: by staying grounded, listening to what your emotions are trying to tell you, and responding in ways that bring you back to center. It's also a key foundation of Big Trust, knowing you can meet discomfort with clarity and calm.

CHAPTER 20

Can I Change What I Feel?

Using Your Emotions for Growth

It was 2 p.m. on October 23, 2018, in Randwick, Sydney. I was about to present at a leadership offsite for Australia's top business supplies company. As the MC read my introduction, a surge of intense energy shot through me. My knees wobbled.

"Please join me in welcoming to the stage Shadé Zahrai."

I stepped out to a room full of senior leaders . . . and froze. Completely. I'd spoken at events as part of my corporate banking role before, but this? This was different. This wasn't *part of my day job.* This was the first step into the speaking career I had dreamed of—three years before my experience with the not-so-young "young chief executives." And right there on that stage, I was paralyzed by the tsunami of nerves crashing through me.

For what felt like an eternity (but was probably ten seconds), I just stood there. Silent. Blank. The clicker felt heavy in my hands. The sharp heat of adrenaline flushed my face. Dozens of eyes stared back, waiting.

Then, out of nowhere, I smiled.

Not the panicked, awkward, *I've-bombed-this* kind of smile, but a full, genuine, ear-to-ear grin. And I heard myself say, "I'm so excited to be here with you all!"

This was an anchoring trick I'd picked up from the world-class pros in the dance world. Whether competing, performing, or teaching workshops, when the nerves hit, you smile. One of the world champions once told me, "Smile through the nerves; your body doesn't know the difference between fear and excitement."

I had used that advice in dance countless times. But never in a high-stakes moment like this.

So right there, I put it to the test. And something clicked.

I started moving, making eye contact, engaging with the audience, connecting. Instead of fighting the adrenaline, I channeled it. The nerves that had me frozen just moments before became a force that energized me.

Growth Mode for Managing Emotions

I've never forgotten the experience of that smile. It's my reminder of just how fast our emotional states can shift—sometimes with just a small change. One moment, you're spiraling into emotional chaos. Then, after a walk, a deep breath, a quick chat with a friend, or a good night's sleep, everything feels different.

The situation doesn't change. *You do.*

When you face any situation that carries risk—rejection, failure, criticism, take your pick—self-doubt creeps in and your brain defaults to seeing it as a big, looming threat you need to defend yourself against. You don't consciously choose this response. It's your survival instinct at work, trying to protect you.

This is your *defense mode* fully activated. Suddenly, your thoughts start spiraling:

What if I'm not enough?

What if I can't handle this?

What if I'm humiliated so badly I have to start over in a new city where no one knows me?

Your brain is preparing you for the apocalypse when really, it's just a new project at work.

In defense mode, your blood vessels constrict, your pupils dilate, and your attention narrows, like a camera lens zooming in on a tiny detail and blurring everything else. It's a survival mechanism, but it also means you're missing out on the bigger picture and any creative solutions that might be right under your nose. Remember that time you tore your place apart looking for your car keys, only to realize they were in your pocket the whole time? Yeah, that's defense mode for you—so fixated on the problem that you can't see the obvious answer.

When you're stuck in this mode, your sole focus is survival, not success. Your brain is too busy defending you to let you perform at your best.

Now imagine the same stressful situation, but this time, instead of seeing it as a threat, you *decide* it's a challenge, a chance to learn, to flex your skills. In that moment, your brain's boardroom (responsible for setting your priorities) doesn't hit the panic button. It starts operating differently.

You're no longer fixated on survival, but seizing the moment. This response is entirely based on whether we *believe* we're up for the challenge, because that changes how we interpret what we're facing.

This is *growth mode.* In this state, your brain's boardroom works with you. Studies show that when you interpret a potentially stressful situation as something manageable or a chance to grow, your body adapts to help you perform. Your body opens up, your circulation improves,[1] and you *literally* see more—your pupils dilate to take in the broader context. Your brain releases neurotrophins, which are proteins involved in neuroplasticity, helping your brain cells grow, adapt, and create new pathways. The sand dunes of your mind, where you keep deepening grooves or make new ones, start to shift. In growth mode, opportunities you couldn't see before suddenly snap into focus.

And here's the wild part.

The physical sensations in both defense and growth modes are *almost identical*. A racing heart, sweaty palms, tense muscles—they're there whether you're scared out of your mind or excited beyond belief. The same pounding heart and butterflies can either mean "I don't have what it takes" right before a final-round job interview, or they could mean "I'm ecstatic" as you sit on the edge of your seat waiting for your favorite band to take the stage.

The difference isn't what's happening to your body—it's how you *interpret* it.

Finding Your Own Smile

When I smiled, both on the dance floor and in front of those leaders, I was tapping into what psychologists call the *facial feedback loop*. A simple act, but powerful enough to shift my entire emotional state.

Science confirms it. A meta-analysis of 138 studies found that your facial expressions do more than just reflect your emotions; they reinforce them. Scowl, and you'll feel more stressed. Smile, and your brain starts leaning toward calm and confidence, even if your nerves are still buzzing underneath.[2]

But I didn't just smile. I reinterpreted what was happening. I felt the nerves, the adrenaline, the racing heart . . . and instead of telling myself "I'm freaking out," I told myself (and the audience), "I'm so excited to be here."

This mental pivot is called *cognitive reappraisal*—reframing how you interpret a feeling in real time. And it's one of the most powerful emotion regulation tools we have.[3]

Harvard researcher Alison Wood Brooks put this to the test. She wanted to know: Can changing how we interpret an emotion actually change how we perform under pressure? So she tossed participants into high-stress situations—public speaking, singing in front of strangers, even taking math exams. But before they began, she asked one group to reframe their anxiety by saying out loud, "I'm excited."

The result? Those who made that tiny shift and reinterpreted what they felt as excitement (rather than anxiety) not only *felt* less anxious but also performed better.[4] And it's not just a one-off. Other studies show that when you choose to see stress as something that can help, not hurt you, you gain sharper focus, stronger motivation, and better follow-through.[5]

What's especially fascinating here is that Brooks didn't ask participants to say, "I'm calm." They didn't try to "zen" their state. And there's a good reason for that. When your body is flooded with adrenaline and cortisol, trying to force yourself into a state of calm doesn't work. Your body knows better.

A more effective move is to work *with* that energy. Don't resist it; harness it. Let it fuel you.

Whether you're cramming for an exam, closing a massive deal, or stepping onto a stage in front of thousands, stress can be your ally—if you let it.

So instead of burning energy trying to suppress what you feel, reframe it. Remind yourself: *I'm ready. I care. My body is preparing me to deliver.*

Let that energy work *for* you, not against you.

PRACTICE: Harness Your Emotional Energy

Next time you notice pre-performance stress or anxiety related to doubting your worthiness, capabilities, or impact, start with a smile. Seriously, try it. It's a quick way to tap into the brain's *feedback loop,* signaling to yourself that you're ready and capable. Then, pause and reframe what you feel. See that surge of energy as your body's way of gearing up to help you excel.

- If your heart races when you think about a career-defining opportunity (that you want), tell yourself, *I'm excited for this step. This is my energy building for success.*
- If you feel butterflies before a high-stakes pitch, think, *I'm focused and ready. My body's giving me exactly what I need to speak with impact.*
- If your muscles tense up over a tough decision, remind yourself, *This is my body sharpening my focus, preparing me to make the best choice.*

And if all else fails? Move. If you're overwhelmed, take that energy somewhere—go for a walk, a jog, or even dance like no one's watching (and hopefully, they aren't). Sometimes, you don't need to think your way through stress. You just need to move through it.

The Hidden Weight of Overload

We need to be real, though. Sometimes, all the reframing, smiling, and moving in the world won't cut it. You're so overloaded, stretched so thin, that you can't even think clearly, let alone reframe what you're feeling. That's because, in those moments, it's often so much more than the emotion itself; it's about the full weight of what you're carrying—emotionally, physically, mentally, and environmentally.

In a packed lecture hall, a psychology professor held up a glass of water.

"How heavy is this glass?" he asks.

The students threw out their guesses—12 ounces, 16 ounces, maybe more.

The professor nodded and then continued: "The actual weight of the glass doesn't matter. What matters is how long I hold it. If I hold it for just a minute, it's light and easy to manage. If I hold it for an hour, my arm will start to ache. But if I hold it all day, my arm will feel numb and paralyzed, and eventually, I'll drop it. The glass hasn't changed—but the longer I hold it, the heavier it feels."

This little story has stuck with me over the years. Because it's not about the glass of water, it's about *everything* we carry—whether it's cognitive, emotional, physical, or sensory. They may start small. Manageable. But hold on to them too long, and they start to wear you down.

Sometimes, what feels like self-doubt *isn't* really self-doubt. It's overload. It's not "I'm not good enough," it's "I'm carrying too much, and my brain is waving a white flag."

There are four hidden "loads" we carry that can amplify our existing vulnerabilities to doubt.

Cognitive load. This is the mental clutter that comes from juggling too many thoughts, decisions, and responsibilities. It's the to-do list that never ends, the racing thought loops that won't let you sleep, the feeling of having twenty-three browser tabs open in your brain—all demanding your attention. Sometimes it's caused by overthinking. Sometimes you just have too much on your plate. Either way, your brain gets overtaxed and then can't switch off.

Emotional load. This is the weight of feelings we carry—like stress, anxiety, frustration—that wear us down. It's not so much the emotions themselves; it's the effort it takes to manage them. Holding it together for others, hiding how you really feel, pretending you're fine when you're not—it all adds up, even if no one sees it.

Physical load. This is the toll from too much work, too little rest, or chronic stress. It leaves you tense, achy, and drained. Even simple tasks feel monumental because you're too tired to trust your capabilities and you're running on empty.

Stimulation load. This is the constant assault of sensory input—texts, pings, scrolling, bright screens, and nonstop noise. Your nervous system goes into overdrive and your focus fries. And it's not just tech. Crowded spaces, too many conversations, or back-to-back meetings can leave you wiped—especially if you're sensitive to stimulation. It can leave you feeling wired, restless, and somehow . . . hollow.

When even *one* of these loads gets too heavy, it makes everything harder. It's no wonder you start doubting yourself or feeling powerless. You're not broken or incapable. You're just *tired.* You're exhausted from carrying too much at once. Your system is completely maxed out.

This is what a lot of advice about self-doubt gets wrong. It tells you to "push through," "hustle harder," "feel the fear and show up anyway." But what it misses is that you can't outthink your way through overload. Sometimes, the bravest thing you can do isn't to charge ahead—it's to step back, reflect, and reassess.

You can choose to keep holding on and letting the weight feel heavier and heavier. Or you can choose to put the load down—to let go of some of the weight, even for a moment.

But how do you know which load to set down? How do you step back enough to see the full picture? You can step onto a (metaphorical) balcony. Let me explain.

Take a Break on Your Balcony

When I first started taking Latin dance classes back in 2009, my group had a tradition: Each Monday we'd head to a local restaurant's salsa night to practice. The dance floor was packed with experienced dancers moving effortlessly to the music. The first time I walked in, I froze. The blur of motion, the chaotic energy—it was too much.

A friend noticed, grabbed my arm, and pulled me up a set of

stairs to a balcony that overlooked the floor. From up there, everything looked different. Calmer. The chaos turned into patterns, and I could watch the pros navigate the space with ease.

That balcony became my safety net. My reset button. Even after I gained confidence on the dance floor, I'd retreat there whenever I needed to catch my breath and see the bigger picture. It gave me perspective.

This idea—stepping back from the action to reassess—isn't only useful for dance. It's essential for life. Leadership experts often talk about the importance of "working *on* the business, not *in* the business," stepping away from the daily grind to see things clearly.

The same applies when self-doubt or anxiety takes over. It's easy to get lost in the noise, unable to tell whether you're just overwhelmed or if something deeper is misaligned.

Stepping back to your *figurative* balcony gives you clarity. It allows you to pause, assess what's driving your emotions, and decide what needs to shift.

In my interviews with high performers, this was a common thread. They weren't anxiety free—they just knew how to step onto their figurative balconies to observe their fears and doubts with detachment. From that more objective view, they could see what was weighing them down and choose which load to set aside, even if only for a while.

Your balcony can take many forms:

- Setting a date to revisit a tough decision instead of rushing it
- Taking a walk to clear your head
- Journaling your chaotic thoughts
- Taking a moment to meditate or pray
- Talking things through with someone you trust

These simple actions create breathing room to step back from the chaos, regain some perspective, and see your next move with fresh eyes.

But there's something else that your balcony gives you, beyond managing overwhelm. It can reveal deeper truths. Not all anxiety is just stress. It could be your intuition trying to tell you something isn't right.

That's why stepping onto your balcony matters. It gives you the mental space to ask: *Is this just discomfort I can work through, or is this a deeper misalignment?* It helps you separate fear from intuition, self-doubt from self-awareness—and gives you the courage to act on what you find.

PRACTICE: Step Onto Your Balcony

Taking time to step onto your figurative balcony helps you create space to see the bigger picture, understand your emotional pulse, and make intentional choices about where to direct your energy. Here's how to practice it:

Step 1: Schedule a "Balcony Break"

Put it on your calendar. Once a week, maybe even daily, carve out a little time to step back. First, you're going to spend ten minutes of your Balcony Break doing nothing. That's right, absolutely *nothing*. Sit back or lie down, close your eyes, or even gaze out the window. Just be. The Italians call it *la dolce far niente* (translation: the sweetness of doing nothing). The Dutch call it *niksen* (translation: to do nothing). This practice helps reset your mental networks, allowing your brain to transition from constant "doing" to simply "being."

Once you've given your mind a chance to rest, it's time to take stock of everything you've got going on. Ask yourself:

- What am I currently *feeling* about my life, work, relationships, and commitments?
- Do I feel fired up or completely fried?

Step 2: Check for Excessive Load

Self-doubt is often amplified by hidden burdens. Take a moment to reflect on the underlying loads:

- **Cognitive Load:** Am I overwhelmed by too many decisions, to-dos, or inner judgments?
- **Emotional Load:** Am I carrying the heavy weight of stress, fear, or frustration that's sapping my confidence?
- **Physical Load:** Am I running on empty—exhausted, tense, or sleep-deprived?
- **Stimulation Load:** Am I overstimulated by a flood of notifications, noise, or interactions?

If you answered yes to any of these, your self-doubt might be less about *you* and more about the weight of everything you're lugging around.

Step 3: Lighten the Load

You don't need to carry it all. Lightening your load won't eliminate self-doubt entirely, but it creates the mental and physical space you need to work through it.

- **Cognitive Load: Declutter Your Mind.** Write down everything swirling in your head to free up cognitive space—even a two-minute brain dump helps. If someone else can handle something, let them do that. You don't need to do it all. Then, step away—go for a walk around the block. It helps regulate your nervous system, clears mental fog, and even boosts creativity.[6]

- **Emotional Load: Set Boundaries.** If something drains you, distracts you, or derails your values? It's a "no." Saying no to what doesn't align is saying yes to what matters (plus it honors the Attribute of Acceptance).
- **Physical Load: Refill Your Tank.** Fuel yourself with something nutritious, drink water, stretch or move your body, and don't underestimate the power of sleep. If you can't get more at night, try to fit in a nap or two. Sleep scientists have found that short naps can recharge your mental batteries and improve focus.[7]
- **Stimulation Load: Quiet the Noise.** Your nervous system is not built for constant input. Unplug with intention—log off, mute the pings, and shut the tabs. Even for just ten minutes. Your mind needs space to process, and your nervous system will thank you.

Step 4: Differentiate Between Growth and Misalignment

Self-doubt often shows up as a blanket of fear or anxiety, but stepping back helps you separate growth opportunities from deeper misalignment. Use this guiding prompt to reflect:

If the stress or anxiety I'm feeling vanished tomorrow, would I still want to pursue this path?

From there, evaluate the root of your feelings:

- **Growth Discomfort:** Does this feel like stretching outside your comfort zone? Can you see potential benefits or growth in sticking with it?
- **Misalignment:** Does this path feel draining or misaligned with your values, goals, or sense of self? Are you here because you want it—or because you *think you should* want it?

Step 5: Take Intentional Action

Once you've lightened the load and assessed the underlying message of your doubt, take steps to realign:

- **When it's a sign of growth:** Channel your energy into preparation and practice. Reframe those feelings as signals that you're stepping into new, exciting territory. Lean into your other Attributes to help.
- **When it's misalignment:** Trust your instincts. Give yourself permission to step back or pivot toward something that feels truer to who you are.

Letting go of what's weighing you down creates space for what truly matters, and for the version of you that's ready to show up fully. By stepping onto your balcony, lightening your load, and taking deliberate steps forward, you're choosing to trust yourself again.

• • • •

The energy inside you—whether it feels like anxiety, fear, stress, or uncertainty—doesn't have to hold you back. You can train yourself to see it differently, to turn anxiety into excitement, hesitation into action, and doubt into readiness. Your body isn't working against you; it's preparing you. When you reframe what you feel and lighten the hidden loads weighing you down, you take back control of the moment. And in doing so, you build the habit of grounding yourself. And *that* is both at the heart of Adaptability and a core part of building Big Trust.

CHAPTER 21

The Gift of Awe

Adaptability in the Here and Now

Adaptability begins with tuning into what you're feeling. It becomes more manageable as soon as you begin to separate your feelings from your identity and practice ways to stay emotionally grounded. But there's also a more profound kind of strength that comes when you allow yourself to transcend your self-doubt. It happens when you allow yourself to be moved—by beauty, nature, acts of courage, or moments of connection. In other words, by awe. This feeling of expansiveness gives your whole mind, body, and nervous system a chance to rest and reset.

Jules seemed like she had it all—a thriving global swimwear brand, a happy family, and a life that looked like it was ripped from an Instagram feed. In just eight years, she'd taken a tiny boutique pop-up and turned it into an international success story. From the outside, it was a dream. But Jules didn't see it that way.

She never got enough sleep, was constantly stressed, and always worried everything was going to fall apart. Her overthinking and indecision were tormenting her.

So, one day during one of our group sessions, I switched gears and asked her a question: "When was the last time you felt joy?"

She thought for a moment, then said, "I can't even tell you . . . probably before I started the business. I used to dabble in photography and go bird-watching."

"What did you enjoy about it?" I asked.

Her face softened. "I loved capturing the beauty of nature, the small details that often go unnoticed. Watching the details of birds in their natural environment, it made me feel connected to something bigger than myself."

Jules had been running on empty for so long that she'd forgotten what it was like to feel alive. I encouraged her to carve out space for those moments again. The next week, she lit up telling the group about her weekend. It was the first time she'd taken her focus off her business since the initial icebreakers at our first meeting. The weather had been perfect, and she'd spent hours at a local nature reserve with her binoculars, watching kingfishers by the water. She shared how she lost track of time and was fully present, not thinking about the next task, the next email, or the next fire to put out. Her tone was lighter, her eyes brighter. For the first time in too long, she felt genuinely excited.

She had rediscovered the power of awe.

Savoring Moments of Awe

This is the essence of savoring moments of awe:
Intentionally notice and linger in experiences

of awe to counter life's stressors, restore emotional balance, and reconnect with something larger than yourself.

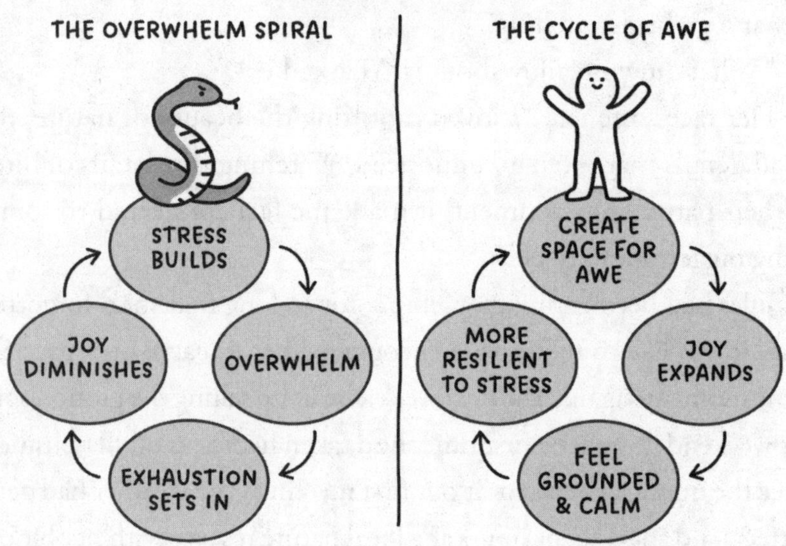

When you're struggling with Adaptability, evoking moments of awe in your life can be your lifeline. Awe has a unique way of lifting you out of your own head and reminding you there's more to life than just deadlines, doubts, and to-do lists.

I travel quite a bit, but one thing I always make time for is spending a little extra time at international arrivals. I can't help it—I'm drawn to emotional reunions.

I remember one time, I was waiting just outside the arrivals gate when I saw an elderly couple walk toward each other. It was obvious they hadn't seen each other in months. Without a word, they fell into each other's arms. I could almost feel their joy and relief from across the room. In that moment, I felt a rush of warmth and emotion. Even though I didn't know them, tears welled up in my eyes. Funny thing, sometimes I'm the only one crying.

Awe, in this case, wasn't dramatic or magnificent. It showed up in the raw, beautiful simplicity of human connection, reminding me of what truly matters.

The "Undoing Effect"

You don't need to climb Everest or backpack through the Amazon to experience awe. Sure, those big, grand adventures are awe-inspiring, but awe is also tucked into the everyday. It's in the micro-moments that don't require a plane ticket or a plan—they're all around you, waiting for you to notice. And as Buddhist nun Pema Chödrön advises, "Rejoicing in ordinary things is not sentimental or trite. It actually takes guts."[1]

Awe happens when we encounter something so powerful we can't wrap our heads around it.[2] Psychologists describe it as a *self-transcendent* or *self-diminishing emotion*—it literally gets you out of your own head. Studies show that when you experience awe, the neural activity tied to overthinking and self-criticism takes a backseat.[3] You're no longer the center of your universe, and guess what? Neither are your problems or your stresses. In this way, awe has a similar effect to the gift of self-forgetting (which helps you tap into your sense of Acceptance).

The "feel good" effects of awe aren't limited to the moment either. It helps reset your entire system. Researchers call this the *undoing effect.*[4] It loosens the grip of heavy emotions like stress,[5] giving your nervous system a chance to recalibrate. You feel lighter. More grounded. More at peace. And that's the beautiful thing about awe. It has the power to momentarily displace your negative state when you're feeling overwhelmed and inject a lasting sense of peace and happiness into your life. These moments are gentle reminders that wonder exists in the smallest of moments, waiting to be noticed, capable of easing our burdens and subtly reshaping our inner world.

Start by noticing what's good around you.[6] Take a walk and look for the everyday details you'd usually miss, big or small: the warmth of sunlight filtering through trees, the sound of birds in the distance, the perfect symmetry of a flower.[7] Reflect on the good things you've done for others or the small wins you've achieved recently.[8] But there is a catch here. Noticing isn't enough. You also have to *savor* the moment.[9]

What does that mean exactly? To savor means to slow down, to fully take in and appreciate what's happening. It's giving yourself permission to feel the joy, the success, the wonder—without downplaying it. Because when you skip this step, you rob yourself of building the belief that you *do* good things, feel good things, and deserve good things.

Here's what savoring moments of awe might look like for you:

- Maybe it's stepping outside at sunset, when the sky melts from peach into lavender, and you stop scrolling. Your mind quiets. You feel the air on your skin. You notice the beauty and it fills you with a quiet kind of reverence . . . and you don't look away.
- Maybe it's witnessing a friend show up for someone else in a hard moment, and something about their generosity moves

you. You let it land. You feel proud to know them. That's also a type of awe.

- Maybe it's sitting in your car after dropping your kids at school, watching them walk away with their oversized backpacks—and feeling a wave of gratitude that they're safe, growing, and okay. You let yourself linger in that moment, instead of rushing to the next thing.

And the best part? Awe makes you more compassionate to *others*.[10] Not only does it help *you* feel better, you become better for the people around you.

Jules took this to heart. She started regularly bird-watching again, even when it felt hard to justify making time for herself. "At first, it was tough to take time to savor these moments because I felt I didn't deserve to," she admitted. "But after I did it a few times, it got easier. I noticed that the more I savored these moments, the more I noticed them, and the more I sought them out. The most remarkable thing was I started worrying less. I became more mindful, more grateful, more patient with my family, and even kinder to myself. I was less in my head and more in the world."

Awe is like an emotional reserve—it fills you up, gives you strength, and makes it easier to face challenges. Over time, you'll likely find that you have a better handle on your self-doubt and the Attributes that fuel it, making room for a more accepting and appreciative view of yourself and your life.

So, take a moment. Step outside. Look up. Pay attention to the little things that make life extraordinary. Because when you build a reservoir of awe, you not only savor the good, you're also cultivating a habit that fundamentally changes the way you show up in the world.

TAKE A MOMENT:

Imagine what you could achieve if you fully embraced Adaptability and savored moments of awe today, this week, and beyond.

THE PATH TO BIG TRUST

CHAPTER 22

Making Self-Trust Your Story

Wherever you are in life right now, your future is still yours to shape. Through this book, you've uncovered *why* self-doubt sticks and learned what to *do* about it. You've learned how you can loosen its grip and build the kind of self-trust that's been within you all along.

As you strengthen who you are at your core, you'll start seeing the patterns more clearly. You'll feel calmer, stronger, more grounded, more like *yourself*. You'll lean on the Attributes that are already strengths, and build the ones that still feel a little shakier.

But remember this: You don't need to change everything all at once. You don't need to become someone else. As you've seen with many people whose stories you've read, all it takes is one shift—one new habit—to begin seeing yourself differently.

In fact, the more specific you are about the burrs of self-doubt that may be tripping you up, the more power you'll have to change your beliefs and start seeing yourself in a new way.

Seven Steps to Big Trust

Here's a seven-step path you can follow—right now—to begin building Big Trust by strengthening your core Attributes: Acceptance, Agency, Autonomy, and Adaptability.

1. **Know where you are.** Start by completing your Doubt Profile. If you haven't done it yet, take a few minutes to answer the twelve questions. Spot which of your Attributes might need the most support—and where self-doubt might be slowing you down without you even realizing it. Don't take your result as a given. Sit with it. Does it match where you feel stuck? Does it highlight something you've been brushing off or underestimating? This step is all about awareness. You can't change what you don't see.
2. **Picture *who* you want to become—without the noise of self-doubt.** Imagine how you'd show up if you operated in your peak state—trusting yourself more and doubting yourself less. What does that version of you look like? How do they speak, decide, lead, live? What does it feel like? Embody it.
3. **Choose an Attribute to strengthen.** Look at the Attribute that's holding you back the most—the one that keeps tripping you up or stopping you from becoming the version of yourself you just pictured. That's your starting point. Focus on it.[1]
4. **Pick one problem to focus on.** Think about a specific challenge tied to the Attribute you've chosen to strengthen.
   A. *Daily Doubt:* Is the problem something that shows up daily, like not speaking your mind or overthinking your emails?

B. *Big Doubt:* Is the problem something bigger, a more significant issue that impacts multiple areas of your life or blocks your longer-term goals, like switching careers?

5. **Tap the "Gifts" for each Attribute to build momentum.**
    - **Acceptance** invites you to embrace your self-worth and focus on contribution over validation by practicing the gift of self-forgetting.
    - **Agency** helps you to trust your abilities and stay out of comparison by channeling the gift of your inner authority.
    - **Autonomy** reminds you to take the wheel of your own life and make the best of what you have with the gift of hope.
    - **Adaptability** guides you to stay grounded during the storm of emotions with the gift of *awe.*
6. **Try one practice (then another).** Select one practice from the Attribute you're focused on (you'll find all the Big Trust Tools on page 313). Try it out for a week or two. Notice what shifts: Are there changes in how you respond to that voice in your head? Are you showing up differently? Self-awareness is powerful, even if transformation doesn't happen all at once.
7. **Find your people.** Self-trust grows faster in good company. Surround yourself with people who believe in you—even when you don't. These are the folks who hold you accountable, cheer you on, and remind you of who you are when doubt gets loud. If you don't have them yet, start looking. One supportive person can make a world of difference.

Remember, building Big Trust and keeping your Attributes strong is a lifelong practice of reinforcing the habits that shape how you see yourself.

As life changes, so will the way these ideas resonate. Keep coming back—you'll uncover something new every time.

Commit to the Journey

Bear in mind too that we live in a society where we demand immediate results. So when growth and progress feel slow, it's tempting to give up. But those are the exact moments when persistence matters most. The purpose of this book is to help you build Big Trust. It's to shift the habits that shape how you see yourself and how you respond when doubt shows up. The goal is to strengthen your core Attributes so that taking bold action, making meaningful changes, and showing up fully becomes your default. For some people, that shift happens quickly. For others, it unfolds more gradually. Both are valid. Both are enough.

The interesting paradox is that while reshaping how you see yourself and working through your vulnerabilities takes effort up front, it ends up giving you *more energy* in the long run. It's the ultimate delayed-return investment. Fayçal and I have seen it over and over—life just feels easier when self-doubt isn't blocking your path. A little effort now for a radically different life? It's a trade-off we'd take any day.

Commit to the process. You might begin with just one small shift—a practice, a habit, a thought you choose to question instead of believe. Then more. Or you may find yourself drawing on each Attribute's gifts in moments you least expect.

Remember: You've done hard things before. Think of the practices, habits, and skills that once felt out of reach but are now second nature. This journey is no different. What feels unfamiliar now will soon become a natural part of who you are.

Your Life Is a Gift

But here's the final, most important piece: *Don't just do this for you.*

Let your growth ripple outward. Let it serve a greater purpose. Share your gifts and talents. Be the leader who uplifts others. The parent who inspires. The friend who brings light. The human who makes the world better just by being more of who they are.

That's what Big Trust unlocks—not just inner trust, but outer impact.

Make *Big Trust* part of who you are.

You're worth that. And so is the world.

Just keep moving forward. Keep choosing yourself.

Because self-trust isn't built in one grand moment. It's built in every small moment where you decide:

I'm not shrinking.

I'm not hiding.

I'm not doubting.

Not this time.

Now go rewrite your story.

BIG TRUST IN RELATIONSHIPS AND LEADERSHIP

While this book focuses on your personal journey, self-doubt doesn't exist in a vacuum—and neither do you. The habits you develop impact not only how you show up for yourself, but also how you show up in relationships, at work, and as a leader.

Your Doubt Profile in Relationships

Self-doubt often shows up in subtle, unspoken ways with those closest to you. A partner pulling away. A friend going quiet. A child hesitating to share. A parent misreading your tone. Self-doubt can create tension, distance, and friction when our inner vulnerabilities surface—and when we don't yet have the tools to work through them together.

That's why nurturing the Four Attributes within your closest relationships matters just as much as developing them within yourself. They help you build more emotional safety, trust, and space to be seen, heard, and supported—on both sides. Here are some small but powerful ways to start.

Building Acceptance Together

- **Listen Up:** Show them they matter by giving your full attention. ("I hear you're feeling unsure about work/school. Tell me more.")

- **Celebrate Character:** Praise who they are, not just what they do. ("Your kindness really stood out today. It always does.")
- **Interrupt Self-Criticism:** Gently challenge their negative self-talk. ("Hey, I don't think that's true at all. You've proven your capability again and again.")

Boosting Agency Together

- **Recognize Effort:** Highlight the process, not just the outcome. ("I see how much effort you're putting into this. It's impressive and it's paying off.")
- **Normalize Trying New Things:** Support experimenting without pressure. ("Let's try this out, even if we're not 100 percent ready. We can always adjust as we go.")
- **Reflect Back Strengths:** Help them see what they bring. ("You have such a great eye for layout and lighting. What you did with the living room redesign completely transformed the space.")

Fostering Autonomy Together

- **Respect Space:** Affirm their need to recharge. ("I know you prefer time alone. Take all the time you need. I'm here when you're ready.")
- **Support Their Decisions:** Show you trust their judgment. ("Whatever you decide, I'm with you.")
- **Ask What They Need:** Instead of jumping into advice or trying to fix things, try asking the LPB Question: "Do you need a (L) listening ear, a (P) problem solver, or a (B) brainstorm buddy?" Let them tell you what they need.

Developing Adaptability Together

- **Lead with Empathy:** Validate their experience. ("I can see this is really hard right now and you're feeling frustrated.")
- **Take a Pause:** When things get tense, model emotional regulation. ("I care about you and your perspective, and I want to respond thoughtfully. Can I take five minutes to process first?")
- **Move Through Stress Together:** Suggest shared recovery strategies. ("Let's get outside for a walk. Some fresh air might help us reset.")

If you'd like to explore how your Doubt Profile shows up in your relationships, and how to nurture Big Trust together, download the companion guide at bigtrustbook.com to get started.

Your Doubt Profile in Leadership

Over the years, I've had the privilege of working with some of the world's most recognized and successful companies—Deloitte, J.P. Morgan, Nestlé, Microsoft, Electrolux, Coca-Cola, Procter & Gamble, LVMH, and more. No matter the industry or level, one question keeps coming up:

"How do I get the best out of my team?"

The research is clear. Whether you're leading a senior exec team, running daily standups with a dev squad, managing a restaurant crew during dinner rush, or coaching your kid's school debate team—it starts with *you*. Your presence, energy, and mindset ripple through everyone else.

But here's what often gets missed: how well your team handles *their* own self-doubt.

Because when people doubt themselves, they pull back. They hesitate. They don't speak up or step up—even in supportive, inclusive environments. And that hesitation limits growth, performance, and potential.

That's where Big Trust comes in. When leaders intentionally cultivate the four core Attributes, they help their people develop real, lasting self-trust—the kind that shows up under pressure and fuels confident action.

And that's how you build teams that are bold, resilient, and high performing—teams grounded in Big Trust. Here are practical ways to start building that kind of culture.

Cultivating Acceptance in Your Team

- **Spotlight Strengths:** Reinforce the habit of recognition. ("You've got a real knack for simplifying the complex. Your input was invaluable in today's meeting.")
- **Champion Openness:** Make it the norm to be real. ("Let's talk openly about what's working and what's not. No judgment here.")
- **Model Self-Compassion:** Share your own failures and emphasize progress over perfection. ("I got that wrong, and here's what it taught me. Let's celebrate effort and learning, not just results.")

Building Agency in Your Team

- **Clarify with a Pre-Mortem:** Set clear goals and conduct a pre-mortem to foresee challenges. ("Let's think ahead about what obstacles we might face in this project and how we'll tackle them.")

- **Reward Initiative:** Build the habit of rewarding action, regardless of the outcome. ("I appreciate that you took the lead on this. Whether it worked out or not, that's the kind of initiative that moves us forward.")
- **Ditch Comparison Culture:** Highlight individual strengths and avoid team members comparing themselves to others. ("You each bring something unique. That's what makes this team powerful.")

Fostering Autonomy in Your Team

- **Promote a Solution Mindset:** Instead of providing answers, prompt them to come up with three solutions. This focuses them on what's within their control. ("What are three ways we could tackle this? What option do you think would be best?")
- **Offer Choices:** Provide options rather than directives. ("We can approach this in a couple of ways—do you prefer plan A or plan B?")
- **Encourage Self-Reflection:** Encourage them to evaluate their own performance and choices. ("What went well for you this quarter? What would you do differently next time?")

Developing Adaptability in Your Team

- **Model Calm Under Pressure:** Show them what staying grounded looks like. ("Last week was a roller coaster. What helped was taking a step back until I was able to think more rationally.")
- **Set the Emotional Climate:** Emotions are contagious, especially yours as a leader. Set the tone by keeping the energy upbeat and encouraging. ("It's been tough, but I've seen how

capable this team is. Let's take a breath, look at what's working, and figure out our next best step, together.")

- **Check the Team's Pulse:** During a daily or weekly check-in, ask everyone to do a quick self-rating by holding up their fingers to show how they're feeling that day. This gives you a nonintrusive way to spot who might need extra support. ("Let's do a quick one-to-five check-in. Hold up the number of fingers that best matches how you're feeling: Five means you're thriving; one means you're struggling.")

If you'd like to explore ways to support your team and build a culture of Big Trust, you can download a companion guide at bigtrustbook.com to get started.

THE BIG TRUST TOOLS

This book offers a range of practical, evidence-based ways to help you build habits that will support your path to Big Trust. The quick-reference guide below gives you a list of key practices and tools you can return to anytime you need a boost, a reset, or a reminder.

The Self-Doubting Mind

1. **Sync Your Boardroom.** If your brain's boardroom feels scattered or stuck, recalibrate. Is your CEO overwhelmed? Pick one priority and align your tasks. Is your Strategist overthinking? Break a plan into three steps and start with step one. Is your Automations Lead exhausted? Simplify a routine with a checklist. Is your Risk Analyst spiraling? Write out the worst case, and how you'd handle it. Small resets help with clarity.
2. **Schedule "Worry Time."** Designate a specific time each day to address your concerns by writing them down as they arise and reviewing them later. This practice helps you manage your thoughts, freeing up your mental energy for the rest of the day.
3. **Invite Your Internal Allies.** When self-doubt hits: (a) Chat with your Fact Checker and ask: *What evidence do I have that this is true? What evidence do I have that this is NOT true? What's a more balanced way to look at this?* (b) Next, shift focus

with your Confidence Consultant and ask: *What would they say about my ability to handle this?*

4. **Reclaim Your Labels.** When you notice negative and limiting labels, consciously replace them with an affirming, growth-oriented statement. For example, change "I'm intense" to "I'm passionate and driven" or "I'm not good enough" to "I'm capable and improving every day."
5. **Create New Tracks—Take an Opposite Action.** When self-doubt urges you to hold back, do the *opposite* of what you feel. If you feel like shrinking, take up space. This helps rewire your conditioned responses and align your actions with how you *want* to feel rather than reinforcing your state.

Acceptance

1. **Standing Up to Your Inner Lies: "Thanks, but No Thanks."** When you're caught in the moment by an Inner Deceiver (Classic Judge, Misguided Protector, Ringmaster, or Neglecter), notice it, and whack it away while saying, *Thanks, but no thanks*. You're disengaging from the thought and creating emotional space.
2. **Talk Back to Your Inner Deceivers.** When an Inner Deceiver's voice clouds your judgment, follow these steps to clear your path: (a) Recognize: Identify the Deceiver (Classic Judge, Misguided Protector, Ringmaster, or Neglecter); (b) Remember: Recall past successes; (c) Reframe: Imagine what advice you would give a loved one if they had this fear; (d) Respond: Acknowledge and reassure your Inner Deceivers: *[Inner Deceiver Name], thanks for caring and trying to help, but I've got this.*

3. **"Stop Should-ing Yourself"—Peel Off a Mask for a Day.** The word *should* often stems from obligation or people-pleasing. Peel back your "mask" for a day, whether that means sharing your true opinions, asking for help, or wearing something that reflects your personality. *If I believe I should __________, then it's time to __________ instead.*
4. **Align with the Person Behind the Mask.** To start to break free from approval-seeking, get to the root of your fear and reconnect with your true self: (a) Identify why you're seeking approval and what fear is driving it; (b) Ask yourself what it is costing you; (c) Act as if you already had approval; (d) What will you regret *not* doing if you keep playing small? When you consistently choose alignment over approval, you stop performing and start living as your whole, worthy self.
5. **Process over Perfection.** Done is often better than perfect. Ditch "perfection" as your goal, and instead aim to "meet spec"—start with the bare minimum to get started. The simplest action becomes the goal.
6. **Honor Your Failures.** When perfectionism strikes, avoid getting stuck in self-criticism with these three steps: (a) Name your judgy thoughts: Objectively assess what happened; (b) Conduct a reality check: Evaluate which behaviors worked for you; (c) Create an excellence statement: Use what went wrong as a launchpad to get better and focus on learning. Failures can be our best teachers—but we have to reflect on them to extract the lessons.
7. **Tap the Gift of Self-Forgetting.** When self-doubt pulls your attention inward, making you hyper-aware of how you're being perceived, redirect your focus to something bigger. Let go of the

need to constantly self-monitor or impress. Focus instead on the people in front of you or the impact you wish to have. This helps reconnect you with your intrinsic worth.

Agency

1. **Reframe Your Imposter Thoughts.** Replace old narratives with more constructive ones: (a) Catch the thought: Notice the imposter narrative; (b) Reframe it: *I don't belong here* becomes *This is an opportunity to grow into the space I've earned.* (c) Say it out loud (or write it down). Bonus tip: Add the word "yet" to the end of any limiting statement to open the door to possibility.
2. **Use "Essence Qualities" to Build Expertise.** If you're facing imposter thoughts and fixating on what you "lack," elevate your sense of Agency by tapping into your essence qualities: (a) Reflect on achievements and identify essence qualities: Call to mind your essence qualities that helped you achieve past successes; (b) Detect your gaps: Identify skills or competencies you objectively lack; (c) Bridge gaps by applying essence qualities: Pair qualities identified in (a) with gaps from (b).
3. **Gratitude Refocus.** Instead of comparing yourself to others, be grateful for what *you* have. This helps you stay grounded and avoid comparison. Finish this sentence: "I am grateful for ______________."
4. **Stay in Your Lane—Prepare, Don't Compare.** To help you stay in your lane and stay focused on your own path, imagine the ideal outcome, as well as potential obstacles and how you'll

handle them: (a) Anticipate obstacles and "unlucky scenarios": Think about every possible challenge you might face; (b) Build your recovery plan: Develop your strategy for how you'll handle each setback; (c) Visualize: Mentally rehearse both obstacles and your responses.

5. **The Daily Dabble.** Set aside ten minutes a day to engage in a playful creative activity where the outcome doesn't matter. Doodle. Play with clay. Strum a guitar. Write the worst poem ever written. Embrace being a beginner and the messiness of learning something new.
6. **Create a "Play Practice" with Side Quests.** Disconnect from the pressure of outcomes by engaging in activities purely for fun: (a) Choose your adventure: Set your side quest; (b) Play: Dive in and experiment; (c) Reflect on the experience: How did it feel before and after? (d) Adjust and refine: Tweak your process for the next quest and do it all again.
7. **Tap the Gift of Inner Authority.** When your Agency is feeling shaky, anchor yourself in what you know to be true. Trust your own skills and your capacity to learn. Your value doesn't come from being the best. It comes from being fully you.

Autonomy

1. **The Cost Reframe.** Stop trading long-term growth for short-term comfort. When hesitation strikes, your brain focuses on escaping discomfort now and ignores the future cost. Flip the script by asking three questions: What do I *gain* if I do this now? What will this *cost* me if I avoid it now? What will this *cost* me

in six months, or in twelve months? Then take one small step forward. That's how momentum starts.

2. **Be a Bison: Microdose Your Hard.** The best way to get better at doing hard things . . . is to keep doing hard things. Boost your tolerance for discomfort and manage risks better by microdosing challenging tasks: (a) Identify your hard: Focus on something you avoid because it feels too difficult; (b) Break it down: Shrink the task into the tiniest steps; (c) Schedule the dose: Choose how often to implement; (d) Track the shift: Notice how you felt before, during, and after; (e) Scale up when ready: Gradually increase the difficulty. This expands your comfort zone and your Luck Surface Area.
3. **Make "I Could" and "I Will" Lists.** When you start feeling wronged or unfairly treated, shift gears: (a) Start with an "I could" list: Write every possible step you *could* take; (b) Then, create your "I will" list: Commit to what you will do.
4. **Rewrite Your Story.** Externalize the weight of your struggles and rewrite your personal narrative to focus on how it has shaped you: (a) Share your current story: Write about a defining moment you had to overcome; (b) Externalize your story: Ask yourself, *What's the story I'd rather be telling? How would I prefer to see myself handling that situation?* to separate your story from your identity and consider the meaning in your challenges; (c) Re-author your story: Reflect on who you were before and after, and how it has shaped you.
5. **Tap the Gift of Hope.** When things feel out of your control, hope helps you shift your focus back to what *is* within your power. Direct your focus to what you can control and take intentional steps forward, driven by a strong belief in your ability to shape a better future.

Adaptability

1. **The Self Check-In.** Boost your emotional adaptability by checking in with yourself: (a) Name it and plot it: Ask, *What am I feeling right now?* then plot it on the matrix; (b) Ask: *Is this emotion serving me?* If the answer is no, ask, *What small shift could help me feel better and move forward? Who or what could support that shift?* Make a commitment and take action.
2. **Uncover the Reason for the Emotion: What's the True Story?** When stress and anxiety hit, it's often due to violated values or self-imposed expectations. (a) Pinpoint the emotion: Identify and name what you're feeling; (b) Explore the trigger: Identify what sparked the emotion; (c) Connect the Attribute: Identify what part of yourself feels under attack; (d) Ask, *What's the story you're telling yourself, and what's the* true *story?* (e) Do something: Decide on a course of action that reaffirms your values. This process helps you understand and address the root of your emotions.
3. **Harness Your Emotional Energy.** Next time you feel pre-performance self-doubt or anxiety, try to reframe it as excitement, focus, or readiness. Tell yourself, *I'm excited for this,* or *My body's giving me the energy to excel,* to turn stress into a performance boost.
4. **Step Onto Your Balcony.** Take regular "Balcony Breaks" to step back, rest, and reflect on your emotional and mental loads: (a) Schedule a Balcony Break: Ask, *What am I currently feeling about my workload and commitments? Do I feel fired up or completely fried?* (b) Check for excessive load: Are you carrying more than you can handle? (c) Lighten the load: Create space; (d) Differentiate between growth and misalignment: Identify

what your self-doubt is telling you; (e) Take intentional action: Realign with what truly matters.

5. **Tap the Gift of Awe.** When you're overwhelmed or stuck in loops of stress, awe gives your system a reset. Intentionally seek and savor moments that move you. Awe restores emotional balance and reconnects you to something greater than yourself.

MY BIG THANKS

This book simply wouldn't exist without the generosity, wisdom, and support of so many people.

Big thanks, from the bottom of my heart, to:

Fayçal, my partner in all things, for trusting me to take the lead on writing this book, for holding down the fort (and the snacks), and for reminding me that I'm never alone in this. Life is a joy with you by my side.

Both our parents, Maryam and Hooman, Mike and Sya, for your love, your values, and the deep well of strength we come from. So much of who we are is because of you.

Janet Goldstein, for being my brilliant partner in getting this book written and completed. You asked the questions that mattered and guided us back when we drifted. I could not have done it without you.

Wendy Sherman, for seeing the potential and advocating for this book with such heart. And to Callie Deitrick, Jenny Meyer, and Heidi Gall for your support and care throughout the entire process.

My incredible circle of pre-readers—Ayman, Julia, Barry, Stephen, Ream, Aakriti, Mo, Anita, Ana Daniela, Asad, and Akie—for reading the draft manuscript, cheering me on, and sending the perfect words at the perfect times. You reminded me of why this matters. And to my wonderful and whip-smart friend Vanessa Van Edwards, big thanks for being such a support and guide through the entire process.

Biz Mitchell, for helping make this book stronger and more impactful. Our publishers, for welcoming this book with such warmth, thoughtfulness, and care. Anna Paustenbach and Olivia Morris, for seeing the spark and saying yes to this idea when it mattered most. Our incredible marketing and support team: Anna Calame, Isabel Yates, Lucy Wingate, Jaini Haria, Aisling O'Toole, Ashley Yepsen, Aly Mostel, Melinda Mullen, Sarah Schoof, Laina Adler, Kaelee Aboud, Madison Du, Lily Crozier. Blake Attwood, for stepping in at just the right moment and helping us make the book tighter. Maya P. Lim for turning my rough mock-ups into such poignant, powerful visuals. Billy Oppenheimer, whose high-value newsletter surfaced many of the stories that found their way into these pages. Mark Fortier, Anne Day, and my thoughtful PR team, for helping amplify this message and guiding the story into the world. And to everyone who helped share the book and spread the word.

Professors Mariano (Pitòsh) Heyden and Kohyar Kiazad, for challenging and supporting me in all the right ways during the five-year PhD journey. The many clients, students, and professionals I've had the privilege to work with over the years, for your stories, your questions, your courage, and your honesty that helped shape the insights in this book. And to all those who participated in our research, for your time, your trust, and your openness—your experiences are woven into every chapter.

And finally, you, dear reader. The biggest thanks to you for picking up this book. Thank you for letting me sit beside you as you question, reflect, and grow. If it leaves you feeling even a little more brave, a little more kind to yourself, or a little more ready to trust yourself and take that next step, you've made everything worth it.

READY TO GO DEEPER?

If you enjoyed this book and you're ready to take things further, we have additional tools, insights, and resources waiting for you that can extend the impact of *Big Trust* in your life and work. Visit:

bigtrustbook.com

You'll find:

- Downloadable resources, guides, and step-by-step action plans to apply the Attributes in real life.
- Bonus content to help you implement the ideas from the book.
- A *Quick Implementation Toolkit* for coaches, mentors, leaders, and practitioners who want to use the Doubt Profile with clients or teams. It includes conversation guides, practical prompts, and reflection templates to help others build Big Trust—one habit at a time.

Bring *Big Trust* to Your Team or Event

If this work resonated with you and you're curious about how Dr. Shadé could bring the lasting impact of *Big Trust* to your organization or event, visit shadezahrai.com/speaking to learn how she partners with teams and events to spark lasting, meaningful change.

NOTES

Epigraph

1. 'Abdu'l-Bahá, *Tablets of 'Abdu'l-Bahá*, Vol. 2 (Wilmette, IL: Bahá'í Publishing Committee, 1915–1916), 404.

Introduction: We Are All More Than Our Self-Doubts

1. My grandparents (and both Fayçal and I) are followers of the Bahá'í Faith, a beautiful religion that celebrates the unity of all faiths and the oneness of humanity. My grandparents fled Iran to escape religious persecution.

Self-Assessment: Your Self-Doubt Profile

1. These numbers are from a global population of 2,210 people who have taken the Doubt Profile assessment as of April 4, 2025.

Chapter 1: Who's Running the Show?

1. This is based on the work of Professor Julius Kuhl, a leading authority in human motivation: Julius Kuhl, Markus Quirin, and Sander L. Koole, "The Functional Architecture of Human Motivation: Personality Systems Interactions Theory." In *Advances in Motivation Science,* vol. 8, pp. 1–62. Elsevier, 2021.
2. Mark R. Leary and Catherine A. Cottrell, "Evolution of the Self, the Need to Belong, and Life in a Delayed-Return Environment." *Psychological Inquiry* 10, no. 3 (1999): 229–232; Mark Leary, *Understanding the Mysteries of Human Behavior.* Great Courses, 2012.
3. Ian Weinstein, "Don't Believe Everything You Think: Cognitive Bias in Legal Decision Making." *Clinical Law Review* 8 (2002): 783.
4. Ralf Schwarzer, "Thought Control of Action: Interfering Self-Doubts." In *Cognitive Interference*, pp. 99–116. Routledge, 2014.
5. Matthew A. Killingsworth and Daniel T. Gilbert, "A Wandering Mind Is an Unhappy Mind." *Science* 330, no. 6006 (2010): 932.
6. This is called "stimulus-control" for worry; Sarah K. McGowan and Evelyn Behar, "A Preliminary Investigation of Stimulus Control Training for Worry: Effects on Anxiety and Insomnia." *Behavior Modification* 37, no. 1 (2013): 90–112.
7. It's called the Ascending Reticular Activating System, which contains the thalamic reticular nucleus; Nicholas Sheridan and Prasanna Tadi, "Neuroanatomy, Thalamic Nuclei." StatPearls

[Internet] (2025); Mauro Maldonato, "The Ascending Reticular Activating System: The Common Root of Consciousness and Attention." In *Recent Advances of Neural Network Models and Applications: Proceedings of the 23rd Workshop of the Italian Neural Networks Society (SIREN), May 23–25, Vietri sul Mare, Salerno, Italy*, pp. 333–344. Springer International Publishing, 2014.

8. Roger Bohn and James E. Short, "Measuring Consumer Information." *International Journal of Communications* 6 (2012): 980–1000.

Chapter 2: Should I Believe Everything I Think?

1. Robert E. Kleck and Angelo Strenta, "Perceptions of the Impact of Negatively Valued Physical Characteristics on Social Interaction." *Journal of Personality and Social Psychology* 39, no. 5 (1980): 861–873.
2. Hester De Boer, Roel J. Bosker, and Margaretha PC van der Werf, "Sustainability of Teacher Expectation Bias Effects on Long-Term Student Performance." *Journal of Educational Psychology* 102, no. 1 (2010): 168–179.
3. Science has shown that belief can impact outcomes: The placebo effect proves that if you expect less pain, your brain reduces pain-processing activity; Tor D. Wager and Lauren Y. Atlas, "The Neuroscience of Placebo Effects: Connecting Context, Learning and Health." *Nature Reviews Neuroscience* 16, no. 7 (2015): 403–418. Patients who believe in a treatment's effectiveness recover faster, regardless of the treatment's actual efficacy; Janet B. Williams, D. Popp, Ken A. Kobak, and Michael Detke, "P-640 - the Power of Expectation Bias." *European Psychiatry.* 27, supp. 1 (2012): 1. This mind-body connection even affects metabolism, as shown in a Stanford study where participants who believed they consumed a high-calorie milkshake experienced a reduction in hunger, despite drinking the same low-calorie drink; Alia J. Crum, William R. Corbin, Kelly D. Brownell, and Peter Salovey, "Mind over Milkshakes: Mindsets, Not Just Nutrients, Determine Ghrelin Response." *Health Psychology* 30, no. 4 (2011): 424–429.
4. Maxwell Maltz, *The New Psycho-Cybernetics: Updated and Expanded.* Tarcher, 2015..
5. Stanford psychologist, researcher, and author Carol Dweck defines a fixed mindset as a belief that abilities and intelligence are static traits that cannot be changed or developed. In contrast, a growth mindset is the understanding that abilities and intelligence can be developed through dedication, hard work, and learning; Carol S. Dweck, *Mindset: The New Psychology of Success*. Ballantine Books, 2013.
6. Michael Gervais, host, *Finding Mastery with Dr. Michael Gervais*, podcast, "How Creativity Can Help Us Unlock Our Potential | Writer of Pixar's *Inside Out*, Meg LeFauve," January 22, 2024, https://findingmastery.com/podcasts/meg-lefauve/.
7. This is known as the process of neuroplasticity; Eberhard Fuchs and Gabriele Flügge, "Adult Neuroplasticity: More Than 40 Years of Research." *Neural plasticity* 2014, no. 1 (2014): 541870.
8. Lawrence Shapiro, *Embodied Cognition*. Routledge, 2019.

Chapter 3: What Is My Doubt Profile Telling Me?

1. Two quantitative surveys were conducted with a combined 4,078 participants, along with 820 qualitative survey responses, 65 in-depth interviews, and 12 case studies.
2. The Doubt Profile builds on research identifying Core Self-Evaluations (CSEs) as a measur-

able form of self-image, made up of four key levers: self-esteem, self-efficacy, locus of control, and neuroticism. These levers shape how individuals perceive their worth, abilities, and sense of control. The Four Attributes of the Doubt Profile—Acceptance, Agency, Autonomy, and Adaptability—translate these psychological traits into actionable areas for strengthening self-trust and reducing the grip of self-doubt. Timothy A. Judge, Edwin A. Locke, Cathy C. Durham, and Avraham N. Kluger, "Dispositional Effects on Job and Life Satisfaction: The Role of Core Evaluations." *Journal of Applied Psychology* 83, no. 1 (1998): 17–34.

3. As of January 20, 2025, the Doubt Profile framework has been tested with 8,882 people globally.
4. In one of our online programs, learners dedicated just 20 minutes a day for 11 days (3.5 hours total) to self-reflection and applying practical tools. They reported tangible changes in how they view themselves and their inner strength in the face of doubt. In one of the studies from my PhD research with 311 participants, we saw a significant increase in their ability to take meaningful action despite self-doubt after just six weeks of applying targeted tools.

Chapter 4: Am I Enough?

1. Based on data from a global population of 2,210 people who have taken the Doubt Profile assessment as of April 4, 2025.
2. This is a form of extremism related to work, and is usually linked to unmet needs and low self-esteem. Robert J. Vallerand and Virginie Paquette, "On Extreme Behavior and Outcomes: The Role of Harmonious and Obsessive Passion." In *The Psychology of Extremism,* pp. 66–95. Routledge, 2021.
3. Gay Hendricks, *The Big Leap: Conquer Your Hidden Fear and Take Life to the Next Level.* HarperCollins, 2009.
4. Richard H. Smith, Caitlin AJ Powell, David JY Combs, and David Ryan Schurtz, "Exploring the When and Why of *Schadenfreude*." *Social and Personality Psychology Compass* 3, no. 4 (2009): 530–546.
5. Sue Gerhardt, *Why Love Matters: How Affection Shapes a Baby's Brain*. Routledge, 2014.
6. Chris R. Fraley, Nathan W. Hudson, Marie E. Heffernan, and Noam Segal, "Are Adult Attachment Styles Categorical or Dimensional? A Taxometric Analysis of General and Relationship-Specific Attachment Orientations." *Journal of Personality and Social Psychology* 109, no. 2 (2015): 354–368.
7. Dario Cvencek, Ružica Brečić, Elizabeth A. Sanders, Dora Gaćeša, David Skala, and Andrew N. Meltzoff, "Am I a Good Person? Academic Correlates of Explicit and Implicit Self-Esteem During Early Childhood." *Child Development* 95, no. 4 (2024): 1047–1062.
8. Robert E. L. Roberts and Vern L. Bengtson, "Relationships with Parents, Self-Esteem, and Psychological Well-Being in Young Adulthood." *Social Psychology Quarterly* 56, no. 4 (1993): 263–277.
9. Sue Gerhardt, *Why Love Matters: How Affection Shapes a Baby's Brain*. Routledge, 2014.
10. Kristin J. Homan, "Secure Attachment and Eudaimonic Well-Being in Late Adulthood: The Mediating Role of Self-Compassion." *Aging & Mental Health* 22, no. 3 (2018): 363–370.
11. Catherine Cozzarelli, Joseph A. Karafa, Nancy L. Collins, and Michael J. Tagler, "Stability and Change in Adult Attachment Styles: Associations with Personal Vulnerabilities, Life

Events, and Global Construals of Self and Others." *Journal of Social and Clinical Psychology* 22, no. 3 (2003): 315–346; Amir Levine and Rachel Heller, *Attached: The New Science of Adult Attachment and How It Can Help You Find—and Keep—Love*. Tarcher, 2010.

12. Martin Pinquart, Christina Feußner, and Lieselotte Ahnert, "Meta-Analytic Evidence for Stability in Attachments from Infancy to Early Adulthood." *Attachment & Human Development* 15, no. 2 (2013): 189–218.
13. Hala Taha, host, *Young and Profiting Podcast,* podcast, "YAPClassic: Dr. Maya Shankar on Influence and the Science of Decision-Making," May 26, 2023, https://www.youngandprofiting.com/yapclassic-dr-maya-shankar-on-influence-and-the-science-of-decision-making/.
14. Maya Shankar, "Why Change Is So Scary—and How to Unlock Its Potential," video, posted July 25, 2023, by TED Conferences, YouTube, https://www.youtube.com/watch?v=Tt0arZN6EBM.
15. Hei Wan Mak, Taiji Noguchi, Jessica K. Bone, Jacques Wels, Qian Gao, Katsunori Kondo, Tami Saito, and Daisy Fancourt, "Hobby Engagement and Mental Wellbeing Among People Aged 65 Years and Older in 16 Countries." *Nature Medicine* 29, no. 9 (2023): 2233–2240.
16. Ciara M. Kelly, Karoline Strauss, John Arnold, and Chris Stride, "The Relationship between Leisure Activities and Psychological Resources That Support a Sustainable Career: The Role of Leisure Seriousness and Work-Leisure Similarity." *Journal of Vocational Behavior* 117 (2020): 103340.
17. Aytug Cagirtekin and Ozgur Tanriverdi, "Social Hobbies Can Increase Self-Esteem and Quality of Life in Female Breast Cancer Patients with Type A Personality Trait: KRATOS Study." *Medical Oncology* 40, no. 1 (2022): 50.
18. Nicole Farmer, Katherine Touchton-Leonard, and Alyson Ross, "Psychosocial Benefits of Cooking Interventions: A Systematic Review." *Health Education & Behavior* 45, no. 2 (2018): 167–180.
19. Wya Feenstra, Jorik Nonnekes, Tahmina Rahimi, Heleen A. Reinders-Messelink, Pieter U. Dijkstra, and Bas R. Bloem, "Dance Classes Improve Self-Esteem and Quality of Life in Persons with Parkinson's Disease." *Journal of Neurology* 269, no. 11 (2022): 5843–5847.
20. Juliana Breines and Serena Chen, "Self-compassion Increases Self-improvement Motivation." *Personality and Social Psychology Bulletin* 38, no. 9 (2012): 1133–1143.

Chapter 5: Should I Believe the Voices in My Head?

1. Tim Ferriss, host, *The Tim Ferriss Show,* podcast, episode 490, "Dr. Jim Loehr on Mental Toughness, Energy Management, and the Power of Journaling, and Olympic Gold Medals," December 30, 2020, https://tim.blog/2020/12/30/jim-loehr-2-transcript/.
2. Arnaud D'Argembeau, Helena Cassol, Christophe Phillips, Evelyne Balteau, Eric Salmon, and Martial Van der Linden, "Brains Creating Stories of Selves: The Neural Basis of Autobiographical Reasoning." *Social Cognitive and Affective Neuroscience* 9, no. 5 (2014): 646–652.
3. My research seemed to loosely align with the "parts perspective" from Internal Family Systems (IFS), a therapeutic model developed by Richard C. Schwartz.
4. Shadé Zahrai, "Master Your Mindset, Overcome Self-Deception, Change Your Life," video, posted March 30, 2022, by TEDx Talks, YouTube, https://www.youtube.com/watch?v=4AzpmZ7AjaQ.

5. Nele Stinckens, Germain Lietaer, and Mia Leijssen, "Working with the Inner Critic: Process Features and Pathways to Change." *Person-Centered & Experiential Psychotherapies* 12, no. 1 (2013): 59–78.
6. Andrew J. Martin and Herbert W. Marsh, "Fear of Failure: Friend or Foe?" *Australian Psychologist* 38, no. 1 (2003): 31–38.
7. Emma Seppälä, *The Happiness Track: How to Apply the Science of Happiness to Accelerate Your Success.* Hachette UK, 2016.
8. Tal Ben-Shahar, *Happier: Learn the Secrets to Daily Joy and Lasting Fulfillment.* McGraw Hill, 2007.
9. "Obsessive passion" is a term coined by Robert J. Vallerand and colleagues in their 2003 paper, and has been found to be linked to a desire for social acceptance or reflecting low self-esteem: Robert J. Vallerand, Céline Blanchard, Genevieve A. Mageau, Richard Koestner, Catherine Ratelle, Maude Léonard, Marylene Gagné, and Josée Marsolais, "Les Passions de L'ame: On Obsessive and Harmonious Passion." *Journal of Personality and Social Psychology* 85, no. 4 (2003): 756–767; Jeffrey M. Pollack, Violet T. Ho, Ernest H. O'Boyle, and Bradley L. Kirkman, "Passion at Work: A Meta-Analysis of Individual Work Outcomes." *Journal of Organizational Behavior* 41, no. 4 (2020): 311–331.
10. Duncan Cramer, "Acceptance and Need for Approval as Moderators of Self-Esteem and Satisfaction with a Romantic Relationship or Closest Friendship." *The Journal of Psychology* 137, no. 5 (2003): 495–505.

Chapter 6: Why Am I Hiding?

1. Andre Agassi, *Open: An Autobiography.* Vintage, 2010.
2. William G. Graziano and Nancy Eisenberg, "Agreeableness: A Dimension of Personality." In *Handbook of Personality Psychology*, pp. 795–824. Academic Press, 1997.
3. Toru Sato, "Sociotropy and Autonomy: The Nature of Vulnerability." *The Journal of Psychology* 137, no. 5 (2003): 447–466.
4. Ralph L. Piedmont, "Social desirability bias." In *Encyclopedia of Quality of Life and Well-Being Research*, pp. 6036–6037. Springer International Publishing, 2014.
5. Mo Gawdat, *Solve for Happy: Engineer Your Path to Joy*. Simon & Schuster, 2017.
6. Phil Stutz, "Talking About the Tools: John Cusack Interviews Phil Stutz," interview by John Cusack, https://www.thetoolsbook.com/blog/talking-about-the-tools-john-cusack-interviews-phil-stutz-part-3.
7. Dar Meshi, Carmel Morawetz, and Hauke R. Heekeren, "Nucleus Accumbens Response to Gains in Reputation for the Self Relative to Gains for Others Predicts Social Media Use." *Frontiers in Human Neuroscience* 7 (2013): 439.
8. Ronda Rousey, *My Fight Your Fight: The First Memoir by the UFC Star*. Simon & Schuster, 2015.
9. The term is widely attributed to psychologist Clayton Barbeau, who used it to describe the cognitive distortion associated with to the word *should*. However, no original source has been confirmed.
10. Tobias Teichert, Vincent P. Ferrera, and Jack Grinband, "Humans Optimize Decision-Making by Delaying Decision Onset." *PloS One* 9, no. 3 (2014): e89638.

11. Bronnie Ware, *The Top Five Regrets of the Dying: A Life Transformed by the Dearly Departing*. Hay House, 2012.
12. This practice is based on principles of cognitive behavioral therapy.

Chapter 7: Do I Have to Be Perfect?

1. Brené Brown, *The Gifts of Imperfection: Let Go of Who You Think You're Supposed to Be and Embrace Who You Are.* Hazelden Publishing, 2022.
2. Brené Brown, *Atlas of the Heart: Mapping Meaningful Connection and the Language of Human Experience.* Random House, 2021.
3. Joachim Stoeber, ed., *The Psychology of Perfectionism: Theory, Research, Applications.* Routledge, 2018.
4. Paul L. Hewitt, Gordon L. Flett, Samuel F. Mikail, David Kealy, and Lisa C. Zhang, "Perfectionism in the Therapeutic Context: The Perfectionism Social Disconnection Model." In *The Psychology of Perfectionism,* pp. 306–330. Routledge, 2017; Simon B. Sherry, Anna L. MacKinnon, Kristin-Lee Fossum, Martin M. Antony, Sherry H. Stewart, Dayna L. Sherry, Logan J. Nealis, and Aislin R. Mushquash, "Perfectionism, Discrepancies, and Depression: Testing the Perfectionism Social Disconnection Model in a Short-Term, Four-Wave Longitudinal Study." *Personality and Individual Differences* 54, no. 6 (2013): 692–697.
5. Alice Moon, Muping Gan, and Clayton R. Critcher, "The Overblown Implications Effect." *Journal of Personality and Social Psychology* 118, no. 4 (2020): 720–742.
6. Kenneth Savitsky, Nicholas Epley, and Thomas Gilovich, "Do Others Judge Us as Harshly as We Think? Overestimating the Impact of Our Failures, Shortcomings, and Mishaps." *Journal of Personality and Social Psychology* 81, no. 1 (2001): 44–56.
7. Those higher on the Perfectionism scale tend to define excellence on other people's terms, prioritizing what others think of them over what they think of themselves. Andrew P. Hill, Howard K. Hall, and Paul R Appleton, "The Relationship Between Multidimensional Perfectionism and Contingencies of Self-Worth." *Personality and Individual Differences* 50, no. 2 (2011): 238–242.
8. Naomi Eisenberger, Matthew D. Lieberman, and Kipling D. Williams, "Does Rejection Hurt? An fMRI Study of Social Exclusion." *Science* 302, no. 5643 (2003): 290–292.
9. James Dyson, "James Dyson: In Praise of Failure," WIRED, April 11, 2011, https://www.wired.co.uk/article/james-dyson-failure.
10. Excellencism is a positive form of perfectionism and is driven by a desire for growth over self-criticism and worries about what others think. Patrick Gaudreau, Benjamin JI Schellenberg, Alexandre Gareau, Kristina Kljajic, and Stéphanie Manoni-Millar, "Because Excellencism Is More Than Good Enough: On the Need to Distinguish the Pursuit of Excellence from the Pursuit of Perfection." *Journal of Personality and Social Psychology* 122, no. 6 (2022): 1117–1145.
11. Kevin Stoltz and Jeffrey S. Ashby, "Perfectionism and Lifestyle: Personality Differences among Adaptive Perfectionists, Maladaptive Perfectionists, and Nonperfectionists." *Journal of Individual Psychology* 63, no. 4 (2007): 414–423.

12. Joachim Stoeber and Julian H. Childs, "The Assessment of Self-Oriented and Socially Prescribed Perfectionism: Subscales Make a Difference." *Journal of Personality Assessment* 92, no. 6 (2010): 577–585.
13. Elizabeth Gilbert, "Thoughts on Writing," ElizabethGilbert.com, accessed May 24, 2025, https://www.elizabethgilbert.com/thoughts-on-writing/.
14. Seth Godin, "Meeting Spec (Doing the Minimum)," *Seth's Blog*, May 31, 2020, https://seths.blog/2020/05/meeting-spec-doing-the-minimum/.
15. Jennifer Cole Wright, Thomas Nadelhoffer, Tyler Perini, Amy Langville, Matthew Echols, and Kelly Venezia, "The Psychological Significance of Humility." *The Journal of Positive Psychology* 12, no. 1 (2017): 3–12.

Chapter 8: The Gift of Self-Forgetting

1. 'Abdu'l-Bahá. *Star of the West*. Vol. XVII, p. 348. Quoted in *Lights of Guidance: A Bahá'í Reference File*, no. 390. Bahá'í Publishing Trust, 1988.
2. Viktor E. Frankl, *Man's Search for Meaning.* Beacon Press, 2014.
3. Larry Dossey, "The Helper's High," *Explore* 14, no. 6 (2018): 393–399.
4. Joseph Campbell and Bill D. Moyers, *The Power of Myth*. Harmony, 1988.

Chapter 9: Can I Handle This?

1. Albert Bandura called this *self-efficacy*; Albert Bandura, "Self-Efficacy: Toward a Unifying Theory of Behavioral Change." *Psychological Review* 84, no. 2 (1977): 191–215.
2. Bowen Li, "Schema Theory in Personal Growth, Culture, and Social Media: A Literature Review." In *Proceedings of the 2024 10th International Conference on Humanities and Social Science Research (ICHSSR 2024)*, p. 208. Springer Nature, 2024.
3. Clarence Ng, "Mathematics Self-Schema, Motivation, and Subject Choice Intention: A Multiphase Investigation." *Journal of Educational Psychology* 113, no. 6 (2021): 1143–1163.
4. This last one is driven by the Inner Deceiver called the Misguided Protector, which also reflects struggles with Acceptance and fear of failure.
5. David Dunning, "The Dunning-Kruger Effect and its Discontents." *Psychologist* 35 (2022): 2–4.
6. David Dunning shared this in an interview on Adam Grant, host, *ReThinking with Adam Grant,* TED Audio Collective, podcast, "Explaining the Dunning-Kruger Effect and Overcoming Overconfidence with David Dunning," July 16, 2024, https://open.spotify.com/episode/5KG3qHPt3rTdaOxptaKnh7?si=c_4teVpiQDe71NA5WKVxZQ.
7. These principles are grounded in Albert Bandura's social cognitive theory and include verbal persuasion—encouragement from others that reinforces your belief in your abilities—and mastery experiences: past successes that serve as proof of your capability to overcome challenges. Julie Waddington, "Self-Efficacy." *ELT Journal* 77, no. 2 (2023): 237–240.
8. Stanislav Dobrev and Jennifer Merluzzi, "Stayers Versus Movers: Social Capital and Early Career Imprinting Among Young Professionals." *Journal of Organizational Behavior* 39, no. 1 (2018): 67–81.

9. Brian D. McNatt, "Ancient Pygmalion Joins Contemporary Management: A Meta-Analysis of the Result." *Journal of Applied Psychology* 85, no. 2 (2000): 314–322.
10. Robert Rosenthal, "Interpersonal Expectancy Effects: A 30-Year Perspective." *Current Directions in Psychological Science* 3, no. 6 (1994): 176–179.
11. I recommend using what's called the "Reflected Best Self Exercise," which has helped over 26,000 people at world-leading universities and Fortune 500 companies to uncover when they are at their best. Accessed online at https://reflectedbestselfexercise.com/about, the exercise is effective at increasing self-efficacy. Noelle Baird, Jennifer L. Robertson, and Matthew J. W. McLarnon, "Looking in the Mirror: Including the Reflected Best Self Exercise in Management Curricula to Increase Students' Interview Self-Efficacy." *Academy of Management Learning & Education* 22, no. 4 (2023): 662–680.
12. Studies have shown that higher self-efficacy is related to achievement across domains, including academic performance, career goals including salary, and job performance; Kate Talsma, Benjamin Schüz, Ralf Schwarzer, and Kimberley Norris, "I Believe, Therefore I Achieve (and Vice Versa): A Meta-Analytic Cross-Lagged Panel Analysis of Self-Efficacy and Academic Performance." *Learning and Individual Differences* 61 (2018): 136–150; Fred C. Lunenburg, "Self-Efficacy in the Workplace: Implications for Motivation and Performance." *International Journal of Management, Business, and Administration* 14, no. 1 (2011): 1–6.

Chapter 10: Am I an Imposter?

1. Leslie Jamison, "Why Everyone Feels Like They're Faking It," *The New Yorker,* February 6, 2023, https://www.newyorker.com/magazine/2023/02/13/the-dubious-rise-of-impostor-syndrome.
2. Pauline Rose Clance and Suzanne Ament Imes, "The Imposter Phenomenon in High Achieving Women: Dynamics and Therapeutic Intervention," *Psychotherapy: Theory, Research & Practice* 15, no. 3 (1978): 241–247.
3. Joon-Ho Chae, Ralph L. Piedmont, Barry K. Estadt, and Robert J. Wicks, "Personological Evaluation of Clance's Imposter Phenomenon Scale in a Korean Sample." *Journal of Personality Assessment* 65, no. 3 (1995): 468–485; Pauline Rose Clance, Debbara Dingman, Susan L. Reviere, and Dianne R. Stober, "Impostor Phenomenon in an Interpersonal/Social Context: Origins and Treatment." *Women & Therapy* 16, no. 4 (1995): 79–96.
4. Suzanne Pollack, "Was Albert Einstein an Imposter?" Henley Business School, March 30, 2021, https://www.henley.ac.uk/news/2021/was-albert-einstein-an-imposter.
5. Anna Parkman, "The Imposter Phenomenon in Higher Education: Incidence and Impact." *Journal of Higher Education Theory & Practice* 16, no. 1 (2016): 51–60; Loretta Neal McGregor, Damon E. Gee, and K. Elizabeth Posey. "I Feel Like a Fraud and It Depresses Me: The Relation Between the Imposter Phenomenon and Depression." *Social Behavior and Personality: An International Journal* 36, no. 1 (2008): 43–48.
6. Adam Grant (@AdamMGrant), "Impostor syndrome is a paradox . . . ," X [formerly Twitter], November 14, 2021, https://twitter.com/AdamMGrant/status/1459894544884015113.
7. This is an example of a cognitive bias called "magnification and minimization," where you magnify flaws and minimize what you do well, distorting how you see yourself. This bias was

first defined by the American psychiatrist and father of cognitive behavioral therapy, Aaron Beck. Aaron Beck, *Depression: Causes and Treatment.* University of Pennsylvania, 1967.

8. Holly M. Hutchins, Lisa M. Penney, and Lisa W. Sublett. "What Imposters Risk at Work: Exploring Imposter Phenomenon, Stress Coping, and Job Outcomes." *Human Resource Development Quarterly* 29, no. 1 (2018): 31–48.
9. Matthew D. Braslow, Jean Guerrettaz, Robert M. Arkin, and Kathryn C. Oleson, "Self-Doubt." *Social and Personality Psychology Compass* 6, no. 6 (2012): 470–482; Edward E. Jones and Steven Berglas, "Control of Attributions About the Self Through Self-Handicapping Strategies: The Appeal of Alcohol and the Role of Underachievement." *Personality and Social Psychology Bulletin* 4, no. 2 (1978): 200–206.
10. Malissa A. Clark, Jesse S. Michel, Ludmila Zhdanova, Shuang Y. Pui, and Boris B. Baltes, "All Work and No Play? A Meta-Analytic Examination of the Correlates and Outcomes of Workaholism." *Journal of Management* 42, no. 7 (2016): 1836–1873.
11. Dax Shepard, host, *Armchair Expert with Dax Shepard*, podcast, episode 291, "Jason Segel," February 8, 2021, https://armchairexpertpod.com/pods/jason-segel.
12. Holly M. Hutchins and Jennifer Flores, "Don't Believe Everything You Think: Applying a Cognitive Processing Therapy Intervention to Disrupting Imposter Phenomenon." *New Horizons in Adult Education and Human Resource Development* 33, no. 4 (2021): 33–47.
13. Carol S. Dweck, "The Power of Yet | TEDxNorrköping," posted September 3, 2014, by TEDx Talks, YouTube, https://www.youtube.com/watch?v=J-swZaKN2Ic.
14. Baslima A. Tewfik, "The Impostor Phenomenon Revisited: Examining the Relationship Between Workplace Impostor Thoughts and Interpersonal Effectiveness at Work." *Academy of Management Journal* 65, no. 3 (2022): 988–1018.
15. "ARTIST SERIES: Paula Scher," 2005, posted January 16, 2011, by Hillman Curtis, Vimeo, https://vimeo.com/18839878?embedded=true&source=vimeo_logo&owner=5595850.
16. "'The Paris Story'–Tinker Hatfield on Air Max 1," posted August 18, 2018, by 43einhalb sneaker store, YouTube, https://www.youtube.com/watch?v=Dr5g132qXAM.
17. Adam Grant refers to these "essence qualities" as "character skills" in his book, *Hidden Potential: The Science of Achieving Greater Things.* Viking, 2023. These were also formally codified in the Values in Action framework in the early 2000s: Christopher Peterson and Martin E. Seligman, "The Values in Action (VIA) Classification of Strengths." In *A Life Worth Living: Contributions to Positive Psychology*. Oxford University Press, 2006.

Chapter 11: Why Is Everyone Better Than Me?

1. E. Thomas Dowd, "Popular Cognitive Therapy." *Journal of Cognitive Psychotherapy* 8, no. 3 (1994): 258.
2. Yongzhan Li, "Upward Social Comparison and Depression in Social Network Settings: The Roles of Envy and Self-Efficacy." *Internet Research* 29, no. 1 (2019): 46–59.
3. Victoria Medvec, Scott F. Madey, and Thomas Gilovich, "When Less Is More: Counterfactual Thinking and Satisfaction Among Olympic Medalists." *Journal of Personality and Social Psychology* 69, no. 4 (1995): 603–610.
4. William M. Hedgcock, Andrea W. Luangrath, and Raelyn Webster, "Counterfactual Thinking and Facial Expressions Among Olympic Medalists: A Conceptual Replication

of Medvec, Madey, and Gilovich's (1995) Findings." *Journal of Experimental Psychology: General* 150, no. 6 (2021): e13–e21.

5. Jerry Seinfeld, "Jerry Seinfeld About Silver Medal," from his comedy performance *I'm Telling You for the Last Time*, posted on January 8, 2009, by burunduk11, YouTube, https://www.youtube.com/watch?v=wAzzCeSXeuY.
6. Nora Rebecca Krott and Gabriele Oettingen, "Mental Contrasting of Counterfactual Fantasies Attenuates Disappointment, Regret, and Resentment." *Motivation and Emotion* 42 (2018): 17–36; Anne Gene Broomhall, Wendy J. Phillips, Donald W. Hine, and Natasha M. Loi, "Upward Counterfactual Thinking and Depression: A Meta-Analysis." *Clinical Psychology Review* 55 (2017): 56–73.
7. Bob Bowman, "Olympics 2012: Michael Phelps Has Mastered the Psychology of Speed," posted June 15, 2012, by *Washington Post,* YouTube, https://www.youtube.com/watch?v=Htw780vHH0o.
8. Sensen Zhang, Yulun Tang, and Shaohong Yong, "The Influence of Gratitude on Pre-Service Teachers' Career Goal Self-Efficacy: Chained Intermediary Analysis of Meaning in Life and Career Calling." *Frontiers in Psychology* 13 (2022): 843276.
9. Thekla Morgenroth, Michelle K. Ryan, and Kim Peters, "The Motivational Theory of Role Modeling: How Role Models Influence Role Aspirants' Goals." *Review of General Psychology* 19, no. 4 (2015): 465–483.
10. Brian Koppelman, host, *The Moment with Brian Koppelman,* podcast, "Aaron Sorkin," February 22, 2023, https://podcasts.apple.com/us/podcast/aaron-sorkin-02-22-23/id814550071?i=1000647135780. See David Marchese, "Aaron Sorkin on How He Would Write the Democratic Primary for 'The West Wing,'" *The New York Times Magazine,* March 2, 2020, https://www.nytimes.com/interactive/2020/03/02/magazine/aaron-sorkin-interview.html.
11. Peter M. Gollwitzer and Paschal Sheeran, "Implementation Intentions and Goal Achievement: A Meta-Analysis of Effects and Processes." *Advances in Experimental Social Psychology* 38 (2006): 69–119.
12. Michael Phelps recounted the incident at a Forbes 30 under 30 summit, reported by Noah Kirsch, "Michael Phelps Talks Growing His Business, Being a Dad and Ryan Lochte," *Forbes,* October 18, 2016, https://www.forbes.com/sites/noahkirsch/2016/10/18/michael-phelps-talks-business-family-and-ryan-lochte/.

Chapter 12: Do I Just Need to Be More Confident?

1. Payam Zamani, *Crossing the Desert: The Power of Embracing Life's Difficult Journeys.* BenBella Books, 2024.
2. Wojciech Zajkowski, Maksymilian Bielecki, and Magdalena Marszał-Wiśniewska, "Are You Confident Enough to Act? Individual Differences in Action Control Are Associated with Post-Decisional Metacognitive Bias." *PLoS One* 17, no. 6 (2022): e0268501.
3. Betty Edwards, *Drawing on the Right Side of the Brain Deluxe: The Definitive.* Penguin, 2012.
4. Ed Catmull and Amy Wallace, *Creativity, Inc.: Overcoming the Unseen Forces That Stand in the Way of True Inspiration.* Random House, 2014.
5. This experiment was initially developed by designer and engineer with Stanford Peter Silk-

man, and later shared by Tom Wujec in his 2010 TED Talk, "Build a Tower, Build a Team," TED Talk, February 2010, https://www.ted.com/talks/tom_wujec_build_a_tower_build_a_team.

6. **Productivity:** Christopher Clifford, Ellis Paulk, Qiyang Lin, Jeanne Cadwallader, Kathy Lubbers, and Leslie D. Frazier, "Relationships Among Adult Playfulness, Stress, and Coping During the COVID-19 Pandemic." *Current Psychology* 43, no. 9 (2022): 8403–8412; **Flow:** Mihaly Csikszentmihalyi. *Flow: The Psychology of Happiness*. Random House, 1992; Joshua Gold and Joseph Ciorciari, "A Review on the Role of the Neuroscience of Flow States in the Modern World." *Behavioral Sciences* 10, no. 9 (2020): 137.
7. Robert Root-Bernstein, Lindsay Allen, Leighanna Beach, Ragini Bhadula, Justin Fast, Chelsea Hosey, Benjamin Kremkow et al. "Arts Foster Scientific Success: Avocations of Nobel, National Academy, Royal Society, and Sigma Xi Members." *Journal of Psychology of Science and Technology* 1, no. 2 (2008): 51–63.
8. Erin C. Westgate and Timothy D. Wilson, "Boring Thoughts and Bored Minds: The MAC Model of Boredom and Cognitive Engagement." *Psychological Review* 125, no. 5 (2018): 689–713; A. Mohammed Abubakar, Hamed Rezapouraghdam, Elaheh Behravesh, and Huda A. Megeirhi, "Burnout or Boreout: A Meta-Analytic Review and Synthesis of Burnout and Boreout Literature in Hospitality and Tourism." *Journal of Hospitality Marketing & Management* 31, no. 8 (2022): 458–503.
9. David Epstein, Range: *Why Generalists Triumph in a Specialized World*. Riverhead Books, 2019.

Chapter 13: The Gift of Inner Authority

1. Donald J. Albers and Constance Reid, "An Interview with George B. Dantzig: The Father of Linear Programming." *The College Mathematics Journal* 17, no. 4 (1986): 292–314; David Mikkelson, "The Legend of the 'Unsolvable Math Problem,'" Snopes, December 3, 1996, https://www.snopes.com/fact-check/the-unsolvable-math-problem/.
2. Ruth Umoh, "Billionaire Richard Branson Reveals Why He's Such a Huge Fan of Always Saying 'Yes,'" CNBC Make It, December 18, 2017, https://www.cnbc.com/2017/12/18/billionaire-richard-branson-reveals-why-he-always-says-yes.html.

Chapter 14: Do My Choices Matter?

1. Some researchers refer to this as autonomy, some refer to it as the locus of control, and others explain it through self-determination theory. They address the same underlying phenomena: that our belief about our ability to exert control is essential for our well-being and success. Lauren A. Leotti, Sheena S. Iyengar, and Kevin N. Ochsner, "Born to Choose: The Origins and Value of the Need for Control." *Trends in Cognitive Sciences* 14, no. 10 (2010): 457–463.
2. Roy F. Baumeister and Mark R. Leary, "The Need to Belong: Desire for Interpersonal Attachments as a Fundamental Human Motivation." In *Interpersonal Development,* 1st ed., p. 33. Routledge, 2007; Christelle Duprez, Véronique Christophe, Bernard Rimé, Anne Congard, and Pascal Antoine, "Motives for the Social Sharing of an Emotional Experience." *Journal of Social and Personal Relationships* 32, no. 6 (2014): 757–787.

3. Brad J. Bushman, "Does Venting Anger Feed or Extinguish the Flame? Catharsis, Rumination, Distraction, Anger, and Aggressive Responding." *Personality and Social Psychology Bulletin* 28, no. 6 (2002): 724–731.
4. Travis Bradberry, "How Complaining Rewires Your Brain for Negativity," *Entrepreneur,* September 9, 2016, https://www.entrepreneur.com/growing-a-business/how-complaining-rewires-your-brain-for-negativity/281734.
5. Stephanie N. L. Schmidt, Joachim Hass, Peter Kirsch, and Daniela Mier, "The Human Mirror Neuron System—A Common Neural Basis for Social Cognition?" *Psychophysiology* 58, no. 5 (2021): e13781.
6. Susan Nolen-Hoeksema and Christopher G. Davis, "'Thanks for Sharing That': Ruminators and Their Social Support Networks." *Journal of Personality and Social Psychology* 77, no. 4 (1999): 801–814.
7. James A. Shepperd and Robert M. Arkin, "Behavioral Other-Enhancement: Strategically Obscuring the Link Between Performance and Evaluation." *Journal of Personality and Social Psychology* 60, no. 1 (1991): 79–88.
8. Qiang Wang, Nathan A. Bowling, and Kevin J. Eschleman, "A Meta-Analytic Examination of Work and General Locus of Control." *Journal of Applied Psychology* 95, no. 4 (2010): 761–768; Matthew D. Braslow, Jean Guerrettaz, Robert M. Arkin, and Kathryn C. Oleson, "Self-Doubt." *Social and Personality Psychology Compass* 6, no. 6 (2012): 470–482.
9. Locus of control was introduced by the father of social learning theory, American psychologist Julian Rotter, in 1966. See also Thomas W. H. Ng, Kelly L. Sorensen, and Lillian T. Eby, "Locus of Control at Work: A Meta-Analysis." *Journal of Organizational Behavior* 27, no. 8 (2006): 1057–1087.
10. Tugba Türk-Kurtça and Metin Kocatürk, "The Role of Childhood Traumas, Emotional Self-Efficacy and Internal-External Locus of Control in Predicting Psychological Resilience." *International Journal of Education and Literacy Studies* 8, no. 3 (2020): 105–115.
11. Tom Brady, "Tom Brady Opens Up—7th Ring Motivation MJ or Belichick | Enemies | Style of Leadership," posted September 20, 2023, by PBD Podcast, YouTube, https://www.youtube.com/watch?v=liz8rZx1NJ8&t=1729s.
12. At the time of writing this book, and at least the most successful in the Super Bowl era: Bryan DeArdo, "Ranking Super Bowl's 20 Best QBs," CBS Sports, February 12, 2024, accessed July 27, 2024 https://www.cbssports.com/nfl/news/ranking-super-bowls-20-best-qbs-patrick-mahomes-near-top-after-third-mvp-win-following-super-bowl-58-victory/.
13. In a comparison of "want-to" and "have-to" goal motivation, "want-to" goals led to greater motivation, a stronger resistance to temptations, and more success in reaching those goals; Marina Milyavskaya, Michael Inzlicht, Nora Hope, and Richard Koestner, "Saying 'No' to Temptation: Want-To Motivation Improves Self-Regulation by Reducing Temptation Rather Than by Increasing Self-Control." *Journal of Personality and Social Psychology* 109, no. 4 (2015): 677–693.
14. Viktor E. Frankl. *Man's Search for Meaning*. Beacon Press, 2014.
15. Thomas W. H. Ng, Kelly L. Sorensen, and Lillian T. Eby, "Locus of Control at Work: A Meta-Analysis." *Journal of Organizational Behavior* 27, no. 8 (2006): 1057–1087.

Chapter 15: Why Can't Life Be Easier?

1. Dick Prouty, Jane Panicucci, and Rufus Collinson. *Adventure Education: Theory and Applications*. Human Kinetics, 2007.
2. Adam Grant, "There's a Name for the Blah You're Feeling: It's Called Languishing." *New York Times,* April 19, 2021, https://www.nytimes.com/2021/04/19/well/mind/covid-mental-health-languishing.html.
3. Teresa Amabile and Steven Kramer, *The Progress Principle: Using Small Wins to Ignite Joy, Engagement, and Creativity at Work.* Harvard Business Review Press, 2011.
4. Jeff M. Martin, "Facing into the Blizzard: Resiliency and Mortality of Native and Domestic North American Ungulates to Extreme Weather Events." *Diversity* 15, no. 1 (2022): 1–11.
5. This is the shortened version of the quote. Derek Moneyberg (@derekmoneyberg), "Marriage is hard. Divorce is hard," Reddit, January 2021, https://www.reddit.com/r/GetMotivated/comments/kzznvv/choose_your_hard_image/; I'm unsure whether Moneyberg is the originator of this quote. Notably, entrepreneur and investor Codie Sanchez has also used the phrase "choose your hard" in her content to emphasize that growth often requires intentional discomfort.
6. James Clear, *Atomic Habits: An Easy & Proven Way to Build Good Habits & Break Bad Ones*. Avery, 2018.
7. Javier A. Granados Samayoa and Russell H. Fazio, "Do I Want to Do This Now? Task Delay as a Function of Valence Weighting Bias." *Personality and Individual Differences* 219 (2024): 112504.
8. Jannica Heinström, *From Fear to Flow.* Chandos Publishing, 2010.
9. Celine Legrand, Christine Naschberger, Yehuda Baruch, and Nikos Bozionelos, "Chance Events in Managers' Careers: Positive and Negative Events, Their Expected and Unexpected Outcomes." *European Management Review* 20, no. 3 (2023): 461–476.
10. The term *Luck Surface Area* was coined by tech entrepreneur Jason Roberts, "How to Increase Your Luck Surface Area," Codus Operandi, 2010, https://www.codusoperandi.com/posts/increasing-your-luck-surface-area.
11. *Desert Island Discs,* podcast, "Christopher Nolan," hosted by BBC, February 2018, https://open.spotify.com/episode/1bGYQH9N1I1qe352i7KXcv?si=Y7UnR2S4TCCTr49SjXTIWw.
12. Lindsey Bahr, "In *Oppenheimer*, Christopher Nolan Builds a Thrilling, Serious Blockbuster for Adults," *AP News,* July 13, 2023, https://apnews.com/article/oppenheimer-christopher-nolan-469cc81e0989f414a20db5508c7630a0.
13. Cristiane Furini, Jociane Myskiw, and Ivan Izquierdo, "The Learning of Fear Extinction." *Neuroscience & Biobehavioral Reviews* 47 (2014): 670–683.
14. This is called systematic desensitization or fear extinction, and it's one of the main ways therapists treat phobias. S. A. Rauch, Afsoon Eftekhari, and Josef I. Ruzek. "Review of exposure therapy: a gold standard for PTSD treatment." *Journal of Rehabilitation Research and Development* 49, no. 5 (2012): 679–687.

Chapter 16: Why Is This Happening to Me?

1. The concept of post-traumatic growth was initially researched and developed by psychologists Richard Tedeschi and Lawrence Calhoun. For a good discussion, see Bibiána Jozefiaková, Natália Kaščáková, Matúš Adamkovič, Jozef Hašto, and Peter Tavel, "Posttraumatic Growth and Its Measurement: A Closer Look at the PTGI's Psychometric Properties and Structure." *Frontiers in Psychology* 13 (2022): 801812.
2. Robert McKee, *Story: Style, Structure, Substance, and the Principles of Screenwriting*. HarperCollins, 1997.
3. Peter Wohlleben, *The Secret Wisdom of Nature*. Greystone Books, 2019.
4. Oguz A. Acar, Murat Tarakci, and Daan Van Knippenberg, "Creativity and Innovation Under Constraints: A Cross-Disciplinary Integrative Review." *Journal of Management* 45, no. 1 (2019): 96–121.
5. Christopher Peterson and Tracy A. Steen, "Optimistic Explanatory Style." In *Handbook of Positive Psychology*, pp. 244–256. Oxford University Press, 2002.
6. The concept of explanatory style was first researched and developed by leading researcher and the father of modern positive psychology, Martin Seligman; Martin E. P. Seligman, "Explanatory Style: Predicting Depression, Achievement, and Health." In *Brief Therapy Approaches to Treating Anxiety and Depression*, pp. 5–32. Routledge, 2013.
7. Daniel Monehin and Audra Diers-Lawson, "A Model of Pragmatic Optimism for More Effective Crisis Leadership." In *Leadership During a Crisis*, pp. 46–68. Routledge, 2024.
8. Zachary Stockill, "'No complaints': An Interview with Pete Best, the Original Drummer of the Beatles," *PopMatters*, August 25, 2014, https://www.popmatters.com/184898-no-complaints-an-interview-with-pete-best-the-original-drummer-of-th-2495626053.html.
9. Augusto Mellado, María Teresa Del Río, Paola Andreucci-Annunziata, and María Elisa Molina, "Psychotherapy Focusing on Dialogical and Narrative Perspectives: A Systematic Review from Qualitative and Mixed-Methods Studies." *Frontiers in Psychology* 15 (2024): 1308131.
10. 'Abdu'l-Bahá, *Star of the West*, 14, no. 1 (April 1923).
11. NOTE: If you're continually blocked by barriers and obstacles (e.g., systemic racism), you can develop what's called *learned helplessness*. This is where the repeated failures or setbacks (experienced either by you or someone like you) lead to a belief that you're powerless to change your situation, so you stop trying altogether. I'm not talking about that here. That requires a lot more work to change the system for greater equity.
12. Narrative therapy is a form of experiential therapy that encourages trauma survivors to separate their identity from their experiences and rewrite their stories to focus on growth and resilience.
13. This is called the Pennebaker Writing Protocol. James W. Pennebaker, "Expressive Writing in Psychological Science." *Perspectives on Psychological Science* 13, no. 2 (2018): 226–229.

Chapter 17: The Gift of Hope

1. Charles Richard Snyder, "Hope Theory: Rainbows in the Mind." *Psychological Inquiry* 13, no. 4 (2002): p. 249–275.

2. Anthony Ray Hinton and Lara Love Hardin, *The Sun Does Shine*. St. Martin's Press, 2018.
3. Nicholas D. Kristof and Sheryl Wudunn, "The Women's Crusade." *The New York Times Magazine,* August 17, 2009, https://www.nytimes.com/2009/08/23/magazine/23Women-t.html.
4. Quoted by The Female Lead in a Twitter post on May 19, 2021; see The Female Lead (@the_female_lead), "'One day you will tell your story of how you've overcome . . .'" Twitter, May 19, 2021, https://x.com/the_female_lead/status/1392392731708862465.

Chapter 18: Can I Manage My Emotions?

1. Based on data from a global population of 2,210 people who have taken the Doubt Profile assessment as of April 4, 2025.
2. Daniel Goleman. *Working with Emotional Intelligence.* Bantam Books, 1998.
3. Based on data from a global population of 2,210 people who have taken the Doubt Profile assessment as of April 4, 2025.
4. Sylvia Chu Lin, Christiane Kehoe, Elena Pozzi, Daniel Liontos, and Sarah Whittle, "Research Review: Child Emotion Regulation Mediates the Association Between Family Factors and Internalizing Symptoms in Children and Adolescents—A Meta-Analysis." *Journal of Child Psychology and Psychiatry* 65, no. 3 (2024): 260–274.
5. Gedolph A. Kohnstamm, Charles F. Halverson Jr., Ivan Mervielde, Valerie L. Havill, and Charles F. Halverson, eds., *Parental Descriptions of Child Personality: Developmental Antecedents of the Big Five?* Psychology Press, 1998.
6. Jonathan Mitchell, "Understanding Meta-Emotions: Prospects for a Perceptualist Account." *Canadian Journal of Philosophy* 50, no. 4 (2020): 505–523.
7. Elaine N. Aron, *The Highly Sensitive Person.* Harmony Books, 1996.
8. Bianca P. Acevedo, Elaine N. Aron, Arthur Aron, Matthew-Donald Sangster, Nancy Collins, and Lucy L. Brown, "The Highly Sensitive Brain: An fMRI Study of Sensory Processing Sensitivity and Response to Others' Emotions." *Brain and Behavior* 4, no. 4 (2014): 580–594.
9. Chiara Van Reyn, Peter Koval, and Brock Bastian, "Sensory Processing Sensitivity and Reactivity to Daily Events." *Social Psychological and Personality Science* 14, no. 6 (2023): 772–783.
10. Elaine N. Aron, *The Highly Sensitive Person.* Harmony Books, 1996.
11. Chunhui Chen, Chuansheng Chen, Robert Moyzis, Hal Stern, Qinghua He, He Li, Jin Li, Bi Zhu, and Qi Dong, "Contributions of Dopamine-Related Genes and Environmental Factors to Highly Sensitive Personality: A Multi-Step Neuronal System-Level Approach." *PloS One* 6, no. 7 (2011): e21636.
12. Deming Wang, Martin S. Hagger, and Nikos L. D. Chatzisarantis, "Ironic Effects of Thought Suppression: A Meta-Analysis." *Perspectives on Psychological Science* 15, no. 3 (2020): 778–793.
13. Ernest H. O'Boyle Jr., Ronald H. Humphrey, Jeffrey M. Pollack, Thomas H. Hawver, and Paul A. Story, "The Relation Between Emotional Intelligence and Job Performance: A Meta-Analysis." *Journal of Organizational Behavior* 32, no. 5 (2011): 788–818.
14. Dana L. Joseph, Jing Jin, Daniel A. Newman, and Ernest H. O'Boyle, "Why Does

Self-Reported Emotional Intelligence Predict Job Performance? A Meta-Analytic Investigation of Mixed EI." *Journal of Applied Psychology* 100, no. 2 (2015): 298–342.

15. Peter Hills and Michael Argyle, "Emotional Stability as a Major Dimension of Happiness." *Personality and Individual Differences* 31, no. 8 (2001): 1357–1364.

Chapter 19: Can I Believe What I Feel?

1. Joe Rogan, host, *The Joe Rogan Experience,* podcast, "1724—Jewel," Spotify, October 25, 2021, https://open.spotify.com/episode/2TRBNGScfO2K3RWRIYJedJ?si=5ac9820e236948dc.
2. Gianpiero Petriglieri, "Emotions Are Data, Too." *Harvard Business Review,* May 9, 2014, https://hbr.org/2014/05/emotions-are-data-too.
3. Susan David, "Emotional Granularity Checklists," Susandavid.com, May 24, 2021, https://www.susandavid.com/resource/emotional-checklist-general/.
4. Mallory J. Feldman, Eliza Bliss-Moreau, and Kristen A. Lindquist, "The Neurobiology of Interoception and Affect." *Trends in Cognitive Sciences* 28, no. 7 (2024): 643–661.
5. Christine D. Wilson-Mendenhall and John D. Dunne, "Cultivating Emotional Granularity." *Frontiers in Psychology* 12 (2021): 703658.
6. Daniel J. Siegel and Tina Payne Bryson, *The Whole-Brain Child: 12 Revolutionary Strategies to Nurture Your Child's Developing Mind.* Delacorte Press, 2011.
7. Matthew D. Lieberman, Naomi I. Eisenberger, Molly J. Crockett, Sabrina M. Tom, Jennifer H. Pfeifer, and Baldwin M. Way, "Putting Feelings into Words," *Psychological Science* 18, no. 5 (2007): 421–428.
8. University of California – Los Angeles, "Putting Feelings into Words Produces Therapeutic Effects in the Brain," *ScienceDaily,* June 22, 2007, www.sciencedaily.com/releases/2007/06/070622090727.htm.
9. Lisa Feldman Barrett. *How Emotions Are Made: The Secret Life of the Brain.* Mariner Books, 2017.
10. June Gruber, Iris B. Mauss, and Maya Tamir, "A Dark Side of Happiness? How, When, and Why Happiness Is Not Always Good." *Perspectives on Psychological Science* 6, no. 3 (2011): 222–233.
11. Mark A. Davis, "Understanding the Relationship Between Mood and Creativity: A Meta-Analysis." *Organizational Behavior and Human Decision Processes* 108, no. 1 (2009): 25–38; Norbert Schwarz, "Situated Cognition and the Wisdom of Feelings: Cognitive Tuning." *The Wisdom in Feelings* 1 (2002): 144–166; Eugénio Oliveira and Luís Sarmento, "Emotional Advantage for Adaptability and Autonomy." In *Proceedings of the Second International Joint Conference on Autonomous Agents and Multiagent Systems,* pp. 305–312. 2003.
12. Norbert Schwarz, "Situated Cognition and the Wisdom of Feelings: Cognitive Tuning." *The Wisdom in Feelings* 1 (2002): 144–166; Eugénio Oliveira and Luís Sarmento, "Emotional Advantage for Adaptability and Autonomy." *In Proceedings of the Second International Joint Conference on Autonomous Agents and Multiagent Systems,* pp. 305–312. 2003.
13. Iris B. Mauss, Nicole S. Savino, Craig L. Anderson, Max Weisbuch, Maya Tamir, and Mark L. Laudenslager, "The Pursuit of Happiness Can Be Lonely." *Emotion* 12, no. 5 (2012): 908–912.
14. Brett Q. Ford and Maya Tamir, "When Getting Angry Is Smart: Emotional Preferences and Emotional Intelligence." *Emotion* 12, no. 4 (2012): 685–689.

Chapter 20: Can I Change What I Feel?

1. Mark D. Seery, "Challenge or Threat? Cardiovascular Indexes of Resilience and Vulnerability to Potential Stress in Humans." *Neuroscience & Biobehavioral Reviews* 35, no. 7 (2011): 1603–1610.
2. Nicholas A. Coles, Jeff T. Larsen, and Heather C. Lench, "A Meta-Analysis of the Facial Feedback Literature: Effects of Facial Feedback on Emotional Experience Are Small and Variable." *Psychological Bulletin* 145, no. 6 (2019): 610–651.
3. Jason T. Buhle, Jennifer A. Silvers, Tor D. Wager, Richard Lopez, Chukwudi Onyemekwu, Hedy Kober, Jochen Weber, and Kevin N. Ochsner, "Cognitive Reappraisal of Emotion: A Meta-Analysis of Human Neuroimaging Studies." *Cerebral Cortex* 24, no. 11 (2014): 2981–2990.
4. Alison Wood Brooks, "Get Excited: Reappraising Pre-Performance Anxiety as Excitement." *Journal of Experimental Psychology: General* 143, no. 3 (2014): 1144–1158.
5. Alia J. Crum, Modupe Akinola, Ashley Martin, and Sean Fath, "The Role of Stress Mindset in Shaping Cognitive, Emotional, and Physiological Responses to Challenging and Threatening Stress." *Anxiety, Stress, & Coping* 30, no. 4 (2017): 379–395; Jeremy P. Jamieson, Alexandra E. Black, Libbey E. Pelaia, Hannah Gravelding, Jonathan Gordils, and Harry T. Reis, "Reappraising Stress Arousal Improves Affective, Neuroendocrine, and Academic Performance Outcomes in Community College Classrooms." *Journal of Experimental Psychology: General* 151, no. 1 (2022): 197–212.
6. Marily Oppezzo and Daniel L. Schwartz, "Give Your Ideas Some Legs: The Positive Effect of Walking on Creative Thinking." *Journal of Experimental Psychology: Learning, Memory, and Cognition* 40, no. 4 (2014): 1142–1152.
7. Sanae Oriyama, Yukiko Miyakoshi, and Toshio Kobayashi, "Effects of Two 15-min Naps on the Subjective Sleepiness, Fatigue and Heart Rate Variability of Night Shift Nurses." *Industrial Health* 52, no. 1 (2014): 25–35.

Chapter 21: The Gift of Awe

1. Pema Chödrön, *The Places That Scare You: A Guide to Fearlessness in Difficult Times.* Shambhala Publications, 2001.
2. Dacher Keltner, *Awe: The New Science of Everyday Wonder and How It Can Transform Your Life.* Penguin Press, 2023.
3. Jennier E. Stellar, Amie M. Gordon, Paul K. Piff, Daniel Cordaro, Craig L. Anderson, Yang Bai, Laura A. Maruskin, and Dacher Keltner, "Self-Transcendent Emotions and Their Social Functions: Compassion, Gratitude, and Awe Bind Us to Others Through Prosociality." *Emotion Review* 9, no. 3 (2017): 200–207.
4. Barbara L. Fredrickson, Roberta A. Mancuso, Christine Branigan, and Michele M. Tugade, "The Undoing Effect of Positive Emotions." *Motivation and Emotion* 24, no. 4 (2000): 237–258.
5. Maria Monroy, Özge Uğurlu, Felicia Zerwas, Rebecca Corona, Dacher Keltner, Jake Eagle, and Michael Amster, "The Influences of Daily Experiences of Awe on Stress, Somatic Health, and Well-Being: A Longitudinal Study During COVID-19." *Scientific Reports* 13, no. 1 (2023): 9336.
6. In an experiment published in 2007, researchers instructed participants to take a daily twenty-minute walk and "try to notice as many positive things around them as they

could . . . and to identify what it was that made it pleasurable." After a week, these participants were happier compared to those who took a standard walk: Fred B. Bryant and Joseph Veroff, *Savoring: A New Model of Positive Experience.* Psychology Press, 2007.

7. Dacher Keltner, *Awe: The New Science of Everyday Wonder and How It Can Transform Your Life*. Penguin Press, 2023.
8. In the Counting Kindness experiment, people who kept track of their acts of kindness for one week were happier that week compared to those who didn't: Keiko Otake, Satoshi Shimai, Junko Tanaka-Matsumi, Kanako Otsui, and Barbara L. Fredrickson, "Happy People Become Happier Through Kindness: A Counting Kindnesses Intervention." *Journal of Happiness Studies* 7, no. 3 (2006): 361–375.
9. Joanne V. Wood, Sara A. Heimpel, and John L. Michela, "Savoring Versus Dampening: Self-Esteem Differences in Regulating Positive Affect." *Journal of Personality and Social Psychology* 85, no. 3 (2003): 566–580.
10. Elias C. Acevedo and Jeremy Tost, "Self-Transcendent Experience and Prosociality: Connecting Dispositional Awe, Compassion, and the Moral Foundations." *Personality and Individual Differences* 214 (2023): 112347.

Chapter 22: Making Self-Trust Your Story

1. This process of oscillating between a positive state (visualizing a desired future) and a negative state (becoming aware of obstacles and current state) is called *affective shifting*, and it increases your focus and ability to reach your goals: Katja M. Friederichs, Nils B. Jostmann, Julius Kuhl, and Nicola Baumann, "The Art of Getting Things Done: Training Affective Shifting Improves Intention Enactment." *Emotion* 23, no. 3 (2023): 651–663.